The Protein Book

A Complete Guide for the Athlete and Coach

Lyle McDonald

This book is not intended for the treatment or prevention of disease, nor as a substitute for medical treatment, nor as an alternative to medical advice. It is a review of scientific evidence presented for information purposes only. Use of the guidelines herein is at the sole choice and risk of the reader.

For information contact:
Lyle McDonald Publishing
PO Box 1713
Salt Lake City, Ut 84110
Email: lylemcd@comcast.net

Cover and interior book design by Jazz Kalsi
Email: jkalsi@gmail.com

ISBN: 978-0-9671456-6-2

FIRST EDITION
FIRST PRINTING

Acknowledgments

First and foremost, I want to thank Jazz Kalsi for both his cover and interior design work. He deserves full credit for making this book look as good as it does.

I want to give special thanks to all of the individuals on my forum who were nice enough to obtain full research papers for me; they facilitated my research on this book greatly.

An extra special thanks goes out to Arthur Harold Smith who went far and beyond the call of duty in helping me to obtain research that I needed.

Finally, thanks to Sarah for her near infinite patience, feedback, and help, as well as putting up with the stacks of research papers all over the office floor during the seemingly endless time that I worked on this project.

Table of Contents

Introduction

If you asked 100 strength and power athletes what the single most important nutrient was, you would get nearly 100 identical answers, a rarity in sports nutrition. That answer would be protein. Now, whether or not this is exactly accurate, there is certainly an element of truth to it.

The word protein even comes from the Greek word 'proteos' meaning 'the first', signifying its primary importance in human nutrition. For while the human body can go indefinitely without carbohydrate intake, and an extremely long time with zero fat intake, insufficient protein for extended periods leads to rapid death.

And while endurance athletes have traditionally ignored protein intake in favor of massive carbohydrate intakes, recent research is identifying important benefits of protein for those types of athletes as well.

The extreme focus of athletes (especially strength/power athletes and bodybuilders) on protein has made them ripe for the kind of hyperbole and false claims that tend to occur in the field of sports nutrition. Protein powders, advertised with claims of INCREASED NITROGEN RETENTION and 357% BETTER PROTEIN SYNTHESIS, are perennial best sellers. Moreso when the advertisement shows a professional bodybuilder, who probably lives on chicken breast and tuna fish just like everybody else, claiming that the powder was the secret of his success.

Yet outside of sports nutrition, when you enter the world of the mainstream, protein seems to be the redheaded stepchild of the nutrition world. Too much is bad for you, gives you kidney problems, destroys your bones, people already eat too much of it and you needn't go out of your way to eat more; I imagine readers have heard it all before.

Between those two extremes, where everyone seems to have a hidden (or not so hidden) agenda, it can be nearly impossible for a well meaning athlete, who simply wants to ensure that they are optimally supporting their training, to determine the truth of the matter. Coaches want to ensure that their athletes training and nutrition are both optimal and run into similar problems regarding the question of protein intake.

My goal in writing this book is to help both athletes and coaches find their way through a minefield of conflicting and often contradictory information. By the time you're finished, even if you can't get anything else right with your sports nutrition, you should be able to ensure that your protein intake in terms of type, amount and timing of intake is not limiting your ability to improve.

Now, you won't find me telling you that a given type or source of protein is singularly the best. Most, if not all, questions in nutrition (or training or supplements for that matter) depend on context; the same is true for training and supplements. The answer to 'What is best?' is almost always 'It depends.' The optimal protein to take around a workout is not the optimal protein to eat during a normal meal; the best choice when calories are at maintenance may not be the best when an athlete is dieting or trying to gain muscle mass, etc. Endurance athletes might conceivably require a different protein to support their training than a strength/power athlete or bodybuilder.

This book covers a tremendous amount of information ranging from basic physiology and digestion to specific application. I'll start with some technical definitions prior to discussing digestion and basic protein metabolism. The next chapters will discuss the issues of protein quality, protein and amino acid requirements. Timing of protein around training is an area of intense research interest and is discussed in some detail; I'll address some of the most common controversies surrounding protein intake as well.

Much of the information in the first half of the book is fairly technical; by the time you're done reading it, you should be able to critically read any advertisements or nutritional claims that you see. If you someone making a claim that goes distinctly against what the research into the topic says, you can probably be safely assured that their motives have more to do with separating you from your money than in helping you succeed as an athlete.

Following the more technically oriented chapters, I next examine whole food proteins; for each I'll discuss where you can find it in food, what advantages or disadvantages it might have for an athlete and topics of that nature. Again, you won't find me saying that any single protein is the best; all proteins have pros and cons and I'll discuss each.

After looking at dietary proteins, I'll examine protein powders which have been a staple of athletic nutrition for decades now. The discussion will be similar to the chapter on whole proteins; I'll look at each protein powder in terms of its pros and cons, along with examining when any given powder might best be used during the day or around training.

In the next chapter, I'll look at some of the currently popular amino acid supplements which may (or may not) have benefits to athletes. Finally, I'll talk about overall application and how to put together all of the previous information depending on the type of sport you're involved in and your goals.

1

Definitions and Basic Background

In this chapter, I want to briefly discuss what protein is, where it is found in the diet, and what it is used for in the body. I'll also discuss the difference between essential and nonessential (also called indispensable and dispensable) amino acids as well as looking at the issue of complete and incomplete proteins.

What is protein?

Proteins are organic compounds made up of carbon, hydrogen, oxygen and nitrogen. The nitrogen part of protein is what not only makes a protein a protein but also sets it apart from both carbohydrate and fats.

Humans can't fix nitrogen from the air like plants. Therefore we need a dietary source of nitrogen; we also have requirements for individual amino acids. Dietary protein provides both. Readers who have heard the term nitrogen balance thrown around may be wondering if this is the nitrogen that is being referred to and the answer is yes. I'll talk about nitrogen balance in Chapter 4 when I discuss protein requirements.

Where is dietary protein found?

With the exceptions of pure sugars and fats, protein is found in some amount in almost all foods, although the amounts can vary drastically. When most people, especially athletes, think of protein foods they probably think of animal source foods such as meat and dairy. Generally speaking, animal source foods provide the most concentrated source of protein.

Red meat, chicken, fish and pork contain essentially no carbohydrate although the fat content can vary from extremely low to extremely high depending on the type and cut of meat. Skinless chicken breast is essentially fat free, containing nothing but protein while a fatty cut of red meat may contain a significant amount of fat along with its protein.

Dairy foods such as milk, cheese and yogurt also contain significant amounts of protein with highly variable amounts of carbohydrate and fat. Full-fat cheese is high in both protein and fat while fat-free cheese is almost pure protein. Milk and yogurt contain carbohydrates in addition to the protein; fat content can vary from high to low (or zero) depending on whether full-fat, low-fat or skim products are chosen.

There are also vegetable sources of proteins with beans (also called legumes) being the primary source; nuts and seeds also contain protein. Fruits and vegetables both contain trace amounts of protein as well.

Since they tend to be a staple of athletic nutrition, I should discuss protein powders and supplements. In the most general terms, protein is available in supplemental form as either protein powder or free form amino acids. Free form amino acids are simply individual amino acids, either by themselves (e.g. L-glutamine or tyrosine) or in some combination.

Some companies now sell products containing powdered essential amino acids (EAAs) or branched chain amino acids (BCAAs) either alone or in combination. Other products containing various mixes of amino acids (either free form or bonded to one another) are also often available.

I should mention that, although food technologies and flavoring are improving by leaps and bounds, free form amino acids tend to be fairly vile tasting. Arginine and ornithine, sold as Growth Hormone (GH) releasers for years, are both disgustingly bad; quite in fact one company sells a separate product meant solely to cover up their taste. A product that was once popular, ornithine keto-glutarate (OKG) has a taste that has been likened to bleach. In contrast, glutamine is very mild and glycine is said to be somewhat sweet.

Readers may be wondering what the L- that comes before most amino acids (e.g. L-leucine, L-glutamine) stands for. Chemically speaking, many molecules in the body come in one of two different shapes, D- or L-. In the human body, only the L-form of nutrients is used; quite in fact, the D-form of some nutrients (such as D-carnitine) can be toxic to the body. In general, throughout this book I'll drop the L- before individual amino acid names.

Protein powders come in three primary forms which are isolates, concentrates and hydrolysates. Protein concentrates typically contain roughly 80% protein with 5-6% carbohydrate and fat while isolates may contain up to 90% protein. Hydrolysates are simply isolates or concentrates which have been pre-digested (digestion of protein is called hydrolysis) by subjecting them to specific enzymes. Practically speaking, you will typically pay the least for a protein concentrate, more for an isolate and the most for a protein hydrolysate. Because of the presence of free form amino acids in protein hydrolysates, they often have a more bitter taste than either concentrates or isolates.

Amino acids: The building blocks of protein

Proteins are made up of individual components called amino acids (AAs) that are attached together in long chains. In the food supply, there are 20 AAs although more occur within the body (1). For example, 3-methylhistidine is an amino acid generated during muscle protein breakdown; hydroxyproline is made from the breakdown of connective tissue.

Individual amino acids are referred to as peptides. When two amino acids are attached, it is called a di-peptide (di means two). Three aminos are a tri-peptide (tri means three) and anything longer than that is usually just referred to as an oligo or polypeptide (oligo means few and poly means many).

Readers may have seen the terms essential amino acids (often abbreviated to EAAs) and inessential amino acids tossed around (the newer terms for this are indispensable and dispensable). In dietary terms, an essential nutrient is one that is not only required by the body for survival but cannot be made by the body; thus it is essential that they are obtained from the diet. In contrast, inessential nutrients can be made within the body (but are still required for health or survival); thus it is not essential that they be obtained from the diet.

Of the 20 total amino acids, 8 are considered essential with the other 12 being inessential. Although I don't want to get into huge detail here, this topic is actually a bit more complicated than I've made it sound. Under specific circumstances, an inessential amino acid can become essential (2) and there are other categories that are sometimes used.

For example, glutamine is normally considered an inessential AA, the body can make sufficient amounts from other sources and it needn't be obtained from the diet. However, under specific conditions such as high stress, trauma, or burn damage, the body may require more glutamine than it can produce. Under those conditions, extra glutamine must come from the diet (typically via supplementation). Thus glutamine is described as being conditionally essential: under certain conditions, it becomes an essential AA.

For the most part, athletes on high protein intakes from quality sources won't need to worry about such details and I'm mentioning it here only for completeness. A list of the essential and inessential AAs appears in Table 1.

Table 1: Inessential and essential amino acids

Inessential	Essential
Alanine	Lysine
Glutamic acid	Isoleucine
Aspartic acid	Leucine
Glycine	Valine
Serine	Threonine
Proline	Methionine
Glutamine	Phenylalanine
Asparagine	Tryptophan
Cysteine	
Tyrosine	
Histidine	
Arginine	

Before moving on, I want to make readers aware of three of the essential AAs: leucine, isoleucine and valine. These are sometimes referred to as the branched chain amino acids (BCAAs) and play a special role in human physiology and muscle growth. I'll discuss the BCAAs in some detail in a later chapter.

What is protein used for in the body

Protein's primary use in the body is structural; that it, protein is used for the production of other substances in the body (1). Most athletes are probably aware that their skeletal muscles are made up of protein but that's not all. Both cardiac (heart) and smooth (surrounding blood vessels and such) muscle are made of protein. Skin and hair both contain a lot of protein as well.

Additionally, many of the hormones in the body (called peptide hormones) are made from protein. This includes insulin, glucagon, growth hormone (GH) and insulin like growth factor (IGF) which are all simply long chains of amino acids linked together. The catecholamine hormones, adrenaline and noradrenaline, are made from the amino acid tyrosine. Albumin, which is used to transport many hormones in the bloodstream, is also made from protein.

However, proteins are not only valuable for structural reasons. Under some conditions, protein can be used to supply energy. For example, a surprisingly large proportion of dietary protein is metabolized in the liver to something called a keto-acid. That keto-acid can be used to make glucose (via a process called gluconeogenesis), ketones or even cholesterol (only one amino acid can be used for this purpose). Keto-acids can also be used to synthesize the inessential amino acids. What gets made in the liver from keto-acids depends to a huge degree on the metabolic state of the body.

Under conditions of low-carbohydrate and/or low-calorie intake, the body will tend to produce either glucose or ketones depending on which amino acid you're looking at. In contrast, with an excess of calories and carbohydrates coming in, the production of cholesterol or even fatty acids could occur from protein. I should note that, while the pathway to convert protein to fat exists, it is almost never utilized to any significant degree.

Dietary proteins and amino acids are turning out to have far more biological effects in the body as well. Recent research has identified bioactive peptides occurring in dietary proteins that can modulate gut function, affect mineral binding, provide anti-bacterial properties, modulate the immune system, exert anti-thrombotic activity, modulate blood pressure by inhibiting angiotensin-converting enzyme (ACE) and that have opioid and other analgesic effects (3). As one specific example, a bioactive present in casein has been shown to exhibit opioid, ACE-inhibitory, and immunomodulatory effects.

In addition to all of the body's uses of protein described above, there are a number of processes of extra importance to athletes. This includes the repair and replacement of damaged proteins, remodeling of the proteins within muscle, bone, tendon and ligaments, maintenance of optimal functioning of all of the metabolic pathways that use amino acids (presumably these pathways are up regulated in athletes due to training), supporting lean body mass gains, supporting immune system function, and possibly others (4). As you'll see in Chapter 4, this makes determination of protein requirements for athletes quite difficult as it is unclear how much additional protein might be required to optimally support all of the processed listed above.

Complete and incomplete proteins

In the not so distant past, proteins were often classified as either complete, meaning that all of the essential AAs were present, or incomplete, meaning that one or more was absent. However, this turns out to be simplistic and incorrect. With a few weird exceptions such as gelatin and collagen, all dietary proteins contain some amount of all of the amino acids, both essential and inessential. In that sense they are all "complete" proteins.

However, this doesn't mean that all dietary proteins are equivalent. Differences in the amino acid profile (along with other aspects of the protein such as digestion speed) affect how well or poorly any given dietary protein will be used by the body to support either maintenance of current body tissues or growth.

Rather than thinking in terms of "complete" or "incomplete", it then becomes more valuable to talk about dietary proteins in terms of having a limiting amino acid. This is defined as the amino acid that is found in the lowest proportion to what is required by the body. The amount of this limiting amino acid will determine, to a great degree, how well that protein is used by the body.

A common example is that grains are low (relative to human requirements) in the amino acid lysine but high in methionine while beans (legumes) are low in methionine and high in lysine. So the limiting amino acid in grains is found in ample supply in legumes and vice versa.

This is where vegetarians first got the idea of combining proteins (for example, a meal of red beans and rice) to get a "complete" protein. By combining foods, which not only differed in their limiting amino acid but also were complementary in terms of what was limiting in one versus what was present in another, a "complete" protein could be obtained.

Frankly, the issue of limiting amino acids becomes a huge issue when small amounts of a single poor quality protein is being consumed; this is common in third world countries and supplementation of individual amino acids can do amazing things to increase protein quality and health. With the types of protein intakes seen in athletes, especially when they come from high quality protein sources, I consider this topic to be a non-issue.

2

Protein Digestion and Absorption

In this chapter, I want to discuss a variety of topics related to the issue of protein digestion, absorption, and subsequent release into the bloodstream. First I want to briefly examine the issue of protein digestion and absorption itself, before moving onto the topic of protein digestion efficiency.

Protein digestion speed, including the issue of fast and slow proteins, is discussed next as that particular topic has been of interest to sports nutrition since the research first started appearing in the late 1990's. The final topic in this chapter addresses the idea that the body can only utilize a fixed amount of protein per meal.

Step 1: Digestion in the mouth, stomach and small intestine

Although mechanical breakdown of protein foods occurs during chewing, almost no actual digestion of proteins occurs in the mouth. Rather, ingested proteins hit the stomach where some digestion and breakdown occurs via hydrochloric acid and the enzyme pepsinogen.

Most protein digestion actually occurs in the small intestine where protein is broken down into smaller and smaller amino acid (AA) chains via a variety of protein digesting enzymes (1). You can think of these enzymes as being like scissors, which cut up the longer chains of protein into smaller chains.

At the end of digestion but prior to absorption, whole proteins have been broken down to produce free form AAs along with some di- and tri-peptides (2 and 3 AA chains respectively); further breakdown and metabolism will occur within the individual cells of the small intestines itself. Protein chains longer than 3 AAs in length are generally not absorbed to a significant degree (2) although extremely small amounts will occasionally "slip" through and reach the bloodstream.

The intestine has a variety of amino acid transporters, each of which can only carry specific amino acids (transporters often carry more than one amino acid). For example, one transporter (called the L transporter) transports leucine along with other neutral amino acids. Another (called the B transporter) transports threonine and the neutral amino acids. The details of the various AA transporter systems aren't that important practically for the most part.

However, since there can be competition between amino acids for transport, it is occasionally theorized or argued that consuming very large amounts of a single amino acid could impair the intake of other AAs. Whether or not this actually occurs with anything but the most extreme diet or supplement regimens is debatable. However, because of the potential for transporter competition, certain AA supplements must often be taken by themselves to avoid competition with other AAs carried by the same transporter.

In any case, while some of those transporters carry single AAs, there are also transporters for di- and tri-peptides. Once again, chains longer than 3 AAs are generally not transported into the bloodstream in any large quantity in normal individuals; they have to be broken down into smaller chains first (3).

Quite in fact, the gut is set up to prevent the absorption of longer chains of AAs for the following reason: when whole or partial proteins get into the bloodstream, it is not uncommon for the body to mount an immune response to that protein. This is one cause of true food allergies and is not uncommon in certain pathological conditions where damage to the gut allows larger protein chains to get into the bloodstream.

The topic of protein digestion has some relevance to certain claims which are sometimes made for sports nutrition supplements, usually involving peptide hormones such as growth hormone (GH) or Insulin-Like Growth Factor 1 (IGF-1). Products are sometimes claimed to contain these hormones, with the implication that they are orally digestible or absorbable. In the first place, both GH and IGF-1 are drugs and extremely expensive. They couldn't legally be placed in a supplement in the first place and, even if they could, the cost would be astronomical.

However, claims are sometimes made are usually along the lines of a given product naturally containing the hormones mentioned above. For example, colostrum, discussed in detail in Chapter 12, contains a small amount of naturally occurring IGF-1 and claims that this can raise blood levels of IGF-1 are often made.

However, looking at the process of protein digestion, it's clear that any orally ingested peptide hormones won't pass through the gut in any significant amounts; they are simply too long to be absorbed. Rather, they will undergo digestion like any other protein and be broken down into very small AA chains (3 AAs or less), generally losing any biological activity they might have had.

The small intestine actually utilizes a fairly large proportion (30-50%) of ingested AAs for protein and hormone synthesis itself. The exact amount retained is determined by the total amount of protein ingested, the quality of the protein ingested, and the presence or absence of other nutrients (1). The different AAs are retained to different degrees as well; for example, the gut will absorb a majority of ingested glutamine for its own use.

The retention of AAs by the gut is thought to "buffer" incoming dietary AAs (1), storing them for later release when food is not available (e.g. between meals or during the night). This helps to prevent large-scale increases in blood AA levels which tend to promote increased AA oxidation and production of urea (a waste product) in the liver. Certain proteins (notably whey and free form AAs) appear to bypass this aspect of gut absporption, leading to rapid and large increases in blood AA levels (with a concomitant increase in AA oxidation in the liver).

What small amount of protein is not absorbed by the small intestine travels to the large intestine, either for metabolism by the bacteria present there or to be excreted in the stool (1). Protein digestion efficiency is discussed below.

Step 2: Liver metabolism

After absorption, metabolism and subsequent release from the small intestine, free form AAs enter the portal vein on the way to the liver. Upon reaching the liver, AAs are metabolized to a great degree; the liver is thought to monitor the body's levels of AAs, adjusting its metabolism accordingly. The liver uses AAs for a number of different purposes including the synthesis of a variety of different proteins; some of those proteins remain in the liver while some are released into the bloodstream (2).

By far and away the primary fate of AAs entering the liver is catabolism (breakdown) for subsequent metabolism. Quite in fact, with the exception of the BCAAs, which are metabolized primarily in muscle, the degrading enzymes for all other AAs are found in highest concentrations in the liver (4); over half of all AAs that reach the liver may be metabolized in the liver (5).

Metabolism in the liver of AAs occurs through two related processes: deamination and transamination. Both reactions start with the removal of the amino group from the amino acid, leaving the carbon skeleton (also called a keto-acid) and ammonium (NH4+); the latter contains the nitrogen component of the AA.

In the case of deamination, the ammonium is formed into waste products such as urea and subsequently excreted in the urine. With transamination, one AA donates its amino group to another compound resulting in the production of a new amino acid and a keto acid. This is how the body synthesizes the inessential AAs.

The keto acids formed by de-/transamination have a number of possible fates in the body depending on the metabolic state. They can be used to produce energy directly in the liver; alternately, they may be used for the synthesis of glucose, fatty acids or ketones (2).

Oxidation and catabolism of AAs in the liver have both an obligatory and regulatory aspect to them. Obligatory losses are those that occur as a consequence of normal body functioning and are considered constant regardless of diet or the body's condition. They will not be discussed further since they cannot be affected by diet or training.

Regulatory losses are those that occur with changes in diet or exercise (exercise primarily affects the metabolism of AAs in the muscle and is discussed in Chapter 3). In terms of diet, feeding has long been known to stimulate AA oxidation in the liver, especially when AAs in excess of requirements are consumed (4); oxidation of individual AAs has also been found to increase or decrease with increasing and decreasing intakes respectively (6-10). The various degrading enzymes up- and down-regulate in response to changing intake.

Quite in fact, the nitrogen balance studies discussed in Chapter 4 often find that protein in excess of requirements is simply oxidized in the liver (11). I should note that, although AA oxidation has a negative connotation (e.g. AAs oxidized in the liver can't be used to support protein synthesis or recovery), this isn't necessarily the case. The increased oxidation of AAs is thought to play a role in the overall "anabolic drive" of the body and may have benefits in terms of increasing muscle mass (12,13).

Step 3: Release into the bloodstream

After metabolism in the liver, AAs are released into the bloodstream where they can be utilized by other tissues such as skeletal muscle, heart, brain and other organs. Between catabolism and being used for protein synthesis, a little less than 25% of the AAs which reached the liver in the first place are actually released into the bloodstream at all (2), most of what is released being the BCAAs.

Even when AAs are infused (delivered directly into the bloodstream via a needle into the vein), 70-75% of the total AAs are absorbed and utilized by the splanchnic bed (liver and gut); the remaining 25-30% are absorbed and used by skeletal muscle (14). In contrast, when BCAAs are infused, skeletal muscle will absorb and use roughly 65-70% of the total (14). Although other tissues utilize them, BCAA are "muscle food" in a very real sense.

Research has found that the increase in plasma AA levels matches the profile of EAAs o the protein ingested but not the profile of nonessential AAs (15). As well, even massive intakes of protein only raise blood concentrations of AAs by a small amount under most circumstances. In one early study subjects were given 3 g/kg of protein, sufficient to double normal AA intake. However, plasma concentrations of most AAs rose only 30% above normal levels, with concentrations of the BCAAs doubling over normal levels (16).

The AAs released into the bloodstream join with AAs being released from other tissues (e.g. muscle protein breakdown) to form an "AA pool", essentially a small store of free AAs that is used to support protein metabolism. I'd note that while each individual tissue has its own free pool, along with the small amount of free AAs in the bloodstream, it's simpler to treat them all as one single combined pool for discussion purposes.

Related to this, it's important to realize that any individual AA is identical to any other of that same type of AA once they have entered the free pool. In fact, unless an AA is labeled radioactively (which is often done for research purposes), it cannot be distinguished from an AA that is already within the body (17). A molecule of leucine from milk is the same as a molecule of leucine from a piece of chicken, which can't be distinguished from a molecule of leucine that was released from skeletal muscle due to protein breakdown.

Conceptually, the free pool provides the link between dietary protein and body protein in that both dietary protein and body protein feeds into the free pool. So AAs coming into the body from the diet are feeding into the pool from one direction while AAs released from muscle protein breakdown are coming into the pool from the opposite direction.

The free AA pool is relatively small, comprising roughly 1% of the total protein stored in the body. In an average 70 kg (154 lb) man, body protein may comprise 10 kg (22 lbs) of AAs; however, the free pool has been estimated to contain only 100 grams of AAs, not including taurine. Only 5 grams of AAs are actually present in the bloodstream. If taurine is included, the size of the free pool increases to 130 grams of AAs (18).

The body appears to maintain the free AA pool within tight limits (19) and measurements under a variety of conditions find extremely similar values for the free AA pool (20, 21). To quote Furst (20) "The fact that intracellular amino acid pattern is reproducible from one individual to another suggests that the concentration of each individual amino acid in the cell is precisely regulated by the biophysical and biochemical mechanisms."

However, as I'll discuss in a later chapter, acute changes in the free AA pool impact significantly on various aspects of metabolism, especially protein synthesis.

Summary

After ingestion, dietary protein undergoes extensive metabolism prior to actually reaching the bloodstream where it can be used by tissues such as skeletal muscle. Digestion occurs to some degree in the stomach prior to extensive digestion in the small intestine; during this process protein is broken down into smaller chains of AAs with free AAs, di- and tri-peptides being the only forms absorbed to any great degree. A large proportion of AAs are retained in the intestine itself, used for protein and hormone synthesis; this is thought to buffer the body against large scale increases in blood AA levels which tend to promote AA oxidation. Retained AAs can be released to the body during periods of no food intake.

After release from the intestine, free form AAs travel to the liver where additional metabolism occurs. While a majority of BCAAs escape the liver into the bloodstream, over half of AAs will be catabolized in the liver and either excreted as waste or used to produce other substances. Various amounts of protein syntheis occurs in the liver as well.

The remain percentage of AAs are then released into the bloodstream where they enter the "free pool" for subsequent use by other tissues in the body. Due to the extensive metabolism that occurs, it's important to note that the AA profile of a given dietary protein only marginally determines the profile of AAs that will be subsequently released into the bloodstream and "seen" (i.e. absorbed and utilized) by skeletal muscle or other tissues. Between tight regulation of the free AA pool and extensive metabolism of dietary proteins in both the gut and liver, the AA profile of the protein ingested has only a small relationship to the AAs released into the bloodstream.

Once in the free pool, there is no difference between a molecule of leucine from chicken and a molecule of leucine from whey protein (or any other protein source). In that sense, if we looked only at the AA profile, there is no difference between 30 grams of protein from chicken breast and 30 grams from a protein powder with the identical AA content.

However, this isn't to say that they will act in the body as identical protein sources. First and foremost, there could be differences in the presence or absence of other nutrients (carbs, fats, vitamins, minerals), a topic I'll discuss in much more detail in later chapters.

Of greater relevance to this chapter, there can be differences in terms of the efficiency or speed of digestion between one protein and another. As it turns out these differences in digestion (primarily in speed) turn out to impact significantly on how a given protein affects whole body metabolism. These issues are discussed next.

Efficiency of protein digestion

Before moving on to the topic of speed of digestion, I'd like to clear up some confusion regarding protein digestibility. It's sometimes claimed in print advertising or articles that a given protein is digested with much higher efficiency than another. Claims are sometimes made that protein powders (especially predigested or hydrolyzed proteins) are digested more efficiently than whole food proteins. Similarly, vegetarian groups sometimes make the claim that vegetable-based proteins are digested with a higher efficiency than animal-based proteins.

Table 1: Digestibility of common protein foods

Food source	Protein digestibility (%)
Egg	97
Milk and cheese	97
Mixed US diet	96
Peanut butter	95
Meat and fish	94
Whole wheat	86
Oatmeal	86
Soybeans	78
Rice	76

Source: National Research Council. Recommended Dietary Allowances, 10th ed. National Academy Press, 1989.

To measure protein digestion efficiency, researchers feed subjects a given amount of protein and then measure how much of that protein comes out the other end (i.e. in the bowel movement). So if 50 grams of protein were fed and 10 grams got pooped out, that would yield a digestion efficiency of 80% (40 grams out of 50 were absorbed). Table 1 lists values for protein digestibility for some commonly eaten whole food proteins.

Clearly animal source proteins are digested with extremely high efficiency in the 95% and higher range; vegetable proteins show far worse digestibility. In terms of daily protein requirements, an increase in total protein intake of 10-20% may be required to offset the decreased digestibility of vegetable protein sources.

Frankly, outside of the clear difference between animal and vegetable based proteins, there's simply no reason to believe that one high-quality protein will digest with significantly greater efficiency than another (22). While I am unaware of digestion efficiency data for protein powders, let's assume that a given product achieved truly 100% efficiency (which is unlikely), that's at most a 3-5% improvement over eggs, meat, dairy protein or even a protein powder with "only" a 95-97% digestibility.

In practical terms, for every 100 grams of protein eaten, an athlete consuming the 100% efficiency protein might get 3-5 more grams of protein into the system, which seems unlikely to significantly impact on growth or recovery. As well, an athlete could simply eat 3-5% more of the other (usually less expensive) protein sources to compensate for the supposed difference.

And while these differences might be important at low levels of intake (i.e. if an individual is only eating 30 grams of low quality protein per day, they would want the highest digestibility possible) they are unlikely to make a significant difference at the intakes typically seen in athletes or recommended in this book. This is even truer given the types of high-quality proteins consumed by athletes.

Despite no real difference in the efficiency of digestion between high-quality proteins, that still doesn't make them identical. As mentioned above, the presence or absence of other nutrients is obviously important; the AA profile of proteins can also vary. Of more relevance to this chapter is the speed of digestion, discussed next.

Speed of digestion

As mentioned in the last chapter, there are basically three forms in which protein can be consumed: whole proteins (this includes food, protein concentrates and isolates), hydrolysates (partially predigested protein powders) and free form amino acids (either in capsule or powdered form). Practically, di- and tri-peptides might be included along with free form AAs. Depending on whom you're talking to, and what they have to sell you, you can find arguments made either for or against either type in terms of it being optimum from a sports nutrition standpoint.

As a general rule, whole proteins will take the longest to digest and start releasing amino acids into the system. Liquid whole proteins, such as protein isolates and concentrates, will digest somewhat faster than solid whole proteins (chicken, beef, etc.) as they are going to be more easily broken down in the stomach (less mechanical breakdown is required).

A recent claim in the supplement world is that protein hydrolysates will provide AAs to the system significantly faster than whole protein powders and isolates. Hence, it's been suggested that consuming protein hydrolysates following training could have benefits in terms of promoting protein synthesis or glycogen storage (23). But is this actually true, are hydrolysates actually absorbed significantly faster than whole proteins (I'll discuss the impact of protein on post-training recovery in Chapter 8)?

A recent study compared the absorption patterns of both casein and whey and their respective hydrolysates (24). Whey and its hydrolysate both raised amino acids essentially identically; the hydrolysate didn't get aminos into the bloodstream any faster. The casein hydrolysate did actually raise blood amino acid levels roughly five to ten minutes faster than the straight casein, and generated a slightly higher peak amino acid level but the difference was small. I consider it unlikely that this difference in digestion speed will impact meaningfully on any aspect of metabolism, protein synthesis, or glycogen storage.

In another study by the same group, whey and pea protein hydrolysates did release amino acids into the bloodstream faster than whole milk (25). However, the milk also contained a significant amount of dietary fat and had a higher caloric density, both of which slow gastric emptying. I don't find the study results either surprising or particularly relevant. It certainly says nothing about the whey or casein hydrolysates being included in commercial post-workout drinks.

Given the higher cost of protein hydrolysates compared to their equivalent concentrate/isolate, on top of the fact that hydrolysates are often quite bitter due to the presence of free form amino acids, I see little reason to choose a hydrolysate over a protein isolate. Assuming it mattered in the first place, the approximately 5-10 minutes difference in digestion speed between casein and its hydrolysate could be offset by simply consuming the casein five to ten minutes sooner. Additionally, there is emerging evidence (discussed in Chapter 8) that slower proteins (or a combination of fast and slow proteins) are superior following training in the first place.

It is true that free form amino acids are absorbed more rapidly than whole proteins, showing a faster increase in blood AA levels and an equally fast decrease (26); whether this is of benefit depends on the context (I'll discuss the topic more below). While it seems logical that individual AAs will be absorbed the most quickly of all protein forms, di- and tri-peptides are actually absorbed fairly quickly themselves, due to the presence of specific transporters for the di- and tri-peptides as mentioned above (2,3). I'll discuss the idea behind EAA supplements in greater detail in Chapter 12.

Fast and slow proteins

It was always more or less implicitly assumed that protein from different sources was absorbed at roughly the same rate. This changed in 1997 when research found that the two different protein fractions found in milk, whey and casein, showed significantly different digestion rates. This was developed into an idea with relevance to both sports nutrition and overall health, that of "fast" and "slow" proteins (27, 28).

In that particular study, 10 males with a normal protein intake were fasted for 10 hours and then given either 43 grams of casein or 30 grams of whey (this was done to equalize the leucine content of the protein meals since whey and casein differ in leucine content). Researchers tracked blood AA levels, along with changes in whole-body protein synthesis and breakdown for hours following consumpton of the protein meals (27). Significant differences were found between whey and casein for all parameters measured.

The primary finding was that whey spiked blood leucine levels, hitting a peak in one hour, returning to baseline in about 4 hours. In contrast, the casein showed a slower rise, hitting a much lower peak around one hour but maintained that same low level of leucine for nearly 7 hours. Figure 1 below shows the pattern of blood leucine concentration after consumption of each protein. I'd note again that both proteins start to hit the bloodstream about the one-hour mark, but they show very different patterns in terms of the magnitude of the peak and duration that blood leucine is sustained for.

The difference in blood AA pattern was attributed primarily to differences in the rate of digestion between proteins; the structure of whey leads to a rapid rate of digestion while casein tends to "clot" in the stomach, delaying digestion by many hours.

In addition, the researchers found that whey protein stimulated protein synthesis (the construction of larger proteins from individual AAs) with no effect on protein breakdown (the breakdown of larger proteins to individual amino acids). Casein had the opposite effect, inhibiting protein breakdown with no effect on protein synthesis. Another observation was that the whey protein stimulated amino acid oxidation to a slightly greater degree than the casein.

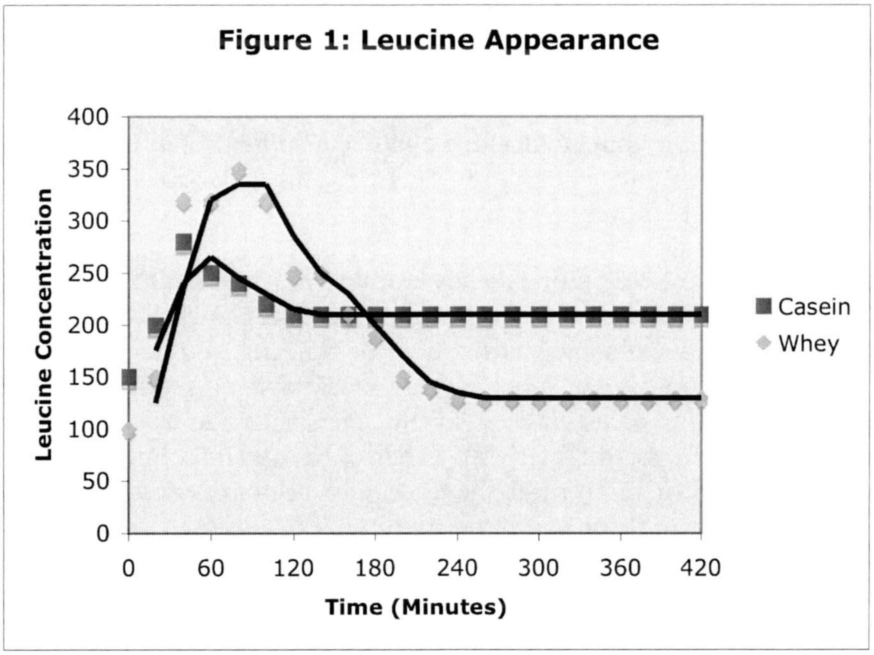

Adapted from: Boirie, Y. et. al. Slow and fast dietary proteins differently modulate postprandial protein accretion. Proc Natl Acad Sci USA (1997) 94: 14930-14935.

Finally, leucine balance, a measure of how much protein was actually retained in the body, was higher for the casein than for the whey. This is an important and often missed point when people examine this research: protein retention, measured in terms of leucine balance, was higher for the casein than whey. This suggests that decreasing protein breakdown, rather than stimulating protein synthesis per se, might have an overall greater impact on the body's protein stores.

Based on this data, casein become known as a slow/anti-catabolic protein and whey was a fast/anabolic protein. The general recommendation became to consume fast proteins like whey around workouts or first thing in the morning (when there was a need to get blood AA levels up rapidly). Casein was recommended at times when an athlete wanted to stave off catabolism or needed a slow rate of digestion (such as before bedtime).

Related to this idea, some authors have argued that a casein/whey blend would impact on both protein breakdown and synthesis. The idea is that the whey will provide a rapid spike of AAs (stimulating protein synthesis) while the casein would maintain a continuous lower level of AAs (inhibiting protein breakdown); there appears to be some merit to this idea.

Preliminary data presented from the 2005 International Whey Conference suggests that a 50/50 mix of whey and casein might provide optimal results in terms of gaining lean body mass with training (29). As above, the idea is that whey provides more of a stimulus to protein synthesis while casein limits protein breakdown.

The combination of the two, a stimulation of protein synthesis with a simultaneous inhibition of protein breakdown, would be expected to have the greatest total impact on body protein stores. Emerging data discussed in Chapter 8 supports this idea. It's interesting to note that bodybuilders have long consumed copious amounts of milk to gain muscle mass; milk contains roughly 80% casein and 20% whey, providing a mixture of both "slow" and "fast" proteins.

But are all of the above conclusions really warranted based on this single study? Some key factors to keep in mind are that the subjects had fasted overnight (which affects protein synthesis and breakdown somewhat), were not training, only a single meal was given to measure the impact on protein synthesis and breakdown (no long-term measure of protein gain was made) and no other nutrients were given with the protein. Thus, it has limited relevance to whey or casein taken with other nutrients, in combination with training, or consumed after one or more meals have already been consumed. Further research has since examined some of those confounding variables.

When casein or whey are taken as part of a mixed meal, the difference between the two in terms of either AA appearance or metabolic effects becomes much less pronounced (30). In that study, while the mixed whey meal released AAs into the bloodstream somewhat more quickly, the mixed casein meal still had the edge on overall protein retention. However, as with the original study, the whey group received less total protein than the casein group; perhaps the casein showed greater total protein gain due to the greater amount of protein ingested, rather than any differences in digestion speed.

In a follow-up study, identical amounts of casein or whey were given with carbohydrate and fat to both young and old individuals (31). The mixed meal containing whey provided a slight benefit for younger individuals compared to casein, and a much greater effect in older people. It's important to note that older individuals appear to respond differently to protein intake than younger people so extrapolation of this data is problematic at best.

Another study found that sipping whey protein made it act more like casein and actually had the greatest impact on protein synthesis compared to either whey or casein given all at once (32). In that study, a total of 30 grams of whey protein was given in 13 discrete drinks (2.3 grams whey protein sipped every 20 minutes) over a period of 4 hours; protein synthesis was higher than in any of the other groups. This type of dosing pattern seems fairly unrealistic for most people. However, it did back up the idea that the big differences between whey and casein were due to differences in digestion speed as mentioned above.

In all of the research cited so far, the researchers were measuring whole body protein synthesis and breakdown, not skeletal muscle synthesis and breakdown specifically. Although popular writers tend to assume that the effect is primarily being seen in skeletal muscle, its just as logical to assume that the proteins being synthesized were in the liver or gut; it turns out that different protein sources do preferentially impact on different tissues in the body.

For example, compared to milk protein, soy tends to be utilized to a greater degree by the gut and liver, providing fewer AAs to peripheral tissues such as muscle (33, 34). This may be due to the speed at which soy is digested (soy is also a fast protein) as well as the overall AA profile; proteins that are low in required AAs may be preferentially utilized by the gut. While maintenance of existing tissues is clearly important, athletes are ultimately concerned with the impact of protein on skeletal muscle in terms of synthesis or breakdown.

Related to this and to the issue of how training might alter the effects of fast versus slow proteins, recently either casein or whey were given one hour after resistance training; both had an identical effect on muscular protein synthesis despite different AA digestion profiles (35).

Another study found that both skim and whole milk consumed after training impacted positively on protein synthesis (with the whole milk having a slightly greater effect) so it may be, as mentioned above, that a mix of casein/whey is superior to either protein individually (36). Related to this, skim milk was recently shown to be superior to soy protein for supporting lean body mass gains with resistance training (37). Further data, discussed in greater detail in Chapter 8 supports the idea that a slow protein, or a mixture of fast/slow proteins following training may be superior to a fast protein alone.

Absorption speed of other proteins

What about the digestion rate of other proteins? Unfortunately not much data exists although one researcher has collected data from a variety of studies to make a series of rough estimations on intestinal protein absorption rates (38). Please note that the numbers below are quite preliminary as some of the studies used fairly indirect measurements to estimate protein digestion rate (I've noted these with an asterisk). A summary of his results appears in Table 2.

Table 2: Estimated intestinal absorption rates of different protein sources

Protein	Absorption rate (g/hour)
Raw egg protein *	1.4
Cooked egg protein *	2.9
Pea protein	3.5
Milk protein	3.5
Soy protein isolate	3.9
Casein isolate	6.1
Whey isolate	8-10
Tenderloin pork steak *	10.0

- These measurements should be considered as the roughest estimates as the studies used indirect measurements of protein digestion.

Clearly there appear to be fairly significant differences in the speed of digestion of different proteins, but there seems to be somewhat of a problem with the above values in real-world terms. Assuming an average digestion rate of roughly six to seven g/h for protein and 24 hours to potentially digest the day's food intake, this would allow for a maximum protein intake of 168 g/day (38). For a 100 kg (220 pound) individual, this would represent a maximum daily protein intake of 1.68 g/kg (0.75 g/lb).

While this is surprisingly consistent with some some estimates of protein requirements discussed in Chapter 4, there are clearly athletes who are consuming far more protein than this on a daily basis. Intake recommendations of 2.5-3.0 g/kg (1.1-1.4 g/lb) and often more are not unusual with some research and researchers supporting that intake level (this is discussed in detail in Chapter 4). Empirically, reports of 300-400 g/day (or more) protein intakes are not uncommon among bodybuilders.

A 100 kg (220 pound) athlete consuming a relatively standard 2.0 g/kg (0.9 g/lb) would still be consuming 200 grams of protein/day, higher than the estimated maximum above. At 3.0 g/kg (1.4 g/lb) or 300 grams of protein per day, he's at nearly double the estimated theoretical maximum based on the digestion rates in Table 2. It seems unlikely that any protein in excess of 168 g/day is simply going undigested.

In a somewhat related vein, studies looking at amino acid infusion suggest that skeletal muscle can handle far more protein than the estimated maximum in terms of stimulating protein synthesis. In one study, protein synthesis rates were examined at different rates of infusion and the maximum response occurred at a rate of 150 mg/kg/h (39). This equates to a dietary intake of 288 g/day or 2.88 g/kg for a 100 kg athlete; this would require a protein digestion speed of roughly 12 g/h over a 24-hour span to achieve. That speed of digestion is higher than any of the values listed in Table 2 above. From a physiological standpoint, it seems highly unlikely that skeletal muscle would be able to respond to a larger amount of protein than can be physiologically digested per day.

Explaining the contradiction

What the above estimations of protein digestion fail to recognize is that the human digestive system can adapt in terms of the rate of gastric emptying (how fast nutrients empty from the stomach) as well as in maximum transport capacity of those nutrients. The gastrointestinal (GI) tract adapts to changes in diet and this effect tends to be nutrient specific (40). Changes in protein intake only affect protein absorption; changes in carb intake only affect carbohydrate absorption, changes in fat intake only affect fat absorption. While much of the data is in animals, there is human data supporting this effect.

One study examined the rate of nutrient transit through the GI tract in active individuals (including several endurance athletes) with a variety of caloric intakes (41). A liquid meal containing 250 calories was given to individuals with drastically differing nutrient intakes. As self-reported caloric intake went up from 1250 to 5300 calories/day, the transit time of the liquid meal dropped from 150-200 minutes to around 50 minutes, a three to four-fold change. The more calories the athletes habitually consumed, the faster they digested their meals.

Studies examining the response to specific nutrients have found similar results. Two weeks of high-fat intake increased the rate of gastric emptying and uptake of dietary fat by about 25% (42). Similarly, three days of high-carbohydrate overfeeding increased the absorption of carbohydrate by about 30% with no impact on the absorption rate of protein (43). No studies specifically looking at protein intake and absorption rates have been done in humans.

However, in rats, a high protein diet fed for 3 weeks increased the rate of gastric emptying of a high protein meal by about 20% (44). This value is at least similar to the changes in humans for carbohydrate and dietary fat so we might expect a similar increase in the rate of protein digestion in humans. Athletes who habitually consume large amounts of protein would be expected to show faster rates of digestion of those proteins than the values shown in Table 2.

The impact of previous meals on digestion speed

Another issue regarding digestion speed that, to my knowledge, has not been examined is the effect of previous meals. As mentioned above, most of the research on this topic (and many others related to protein metabolism) is done in the fasted state. While this serves to minimize the number of variables involved, it raises questions about the real-world applicability of the results.

Meals do not digest immediately and food from a previous meal may be present and still digesting a number of hours later depending on a number of variables such as the size of the meal, form of the meal, macronutrient content, etc. Except for the meal consumed first thing in the morning, all subsequent meals during the day are likely to overlap with the previous meal in terms of digestion; how this impacts on the topics discussed in this chapter is currently unknown. I'll discuss the issue of meal timing and frequency in Chapter 7.

Is faster digestion better?

Before leaving the topic of fast and slow proteins, a question that should be asked is whether a fast speed of digestion is necessarily superior, as this has been the most common interpretation of the studies discussed above (38). It's also conceivably possible that whether a "fast" or "slow" protein is optimal depends on the time of day or condition under which it's consumed. Unfortunately, from a research perspective, there are more questions than answers regarding this topic.

From a practical standpoint, rapidly digesting proteins are generally optimal before or during workouts. With regards to post-workout, it would appear that both slow and fast proteins are equally effective in promoting protein gain, with emerging research suggesting that slower proteins or a mixture of slow and fast might be superior post-workout. This topic is discussed in Chapter 8.

But what about other times of the day? Are slow proteins superior as suggested by the casein versus whey study? Does it matter unless you're consuming protein by itself in the fasted state? Would a mix of fast and slow proteins be better than either individually, or would some other intake pattern give optimal results? Meal frequency is addressed in some detail in Chapter 7 with the issue of protein around training discussed in detail in Chapter 8 and these issues will be addressed in greater detail there.

Is there a maximum protein intake per meal?

As a final issue regarding protein digestion, I want to address a commonly held belief, which is that the body can only handle (assimilate/digest/utilize) a certain amount of protein per meal. Thirty grams is typically given as the magic number but other values are used as well although it's often unclear what exactly a limit is being placed on.

As discussed above, the body appears to increase both the rate of gastric emptying and nutrient absorption in response to increased nutrient intakes and it seems highly unlikely that any single fixed amount of protein could possibly apply to all individuals regardless of body size, activity or habitual protein intake.

If we assume a meal schedule of 6 meals/day (which is fairly standard among bodybuilders), a maximum protein intake of 30 grams per meal would allow for a total of 180 g/day to be consumed. This is simply not consistent with real world intakes among athletes.

Larger athletes often eat far more protein than this and they don't all increase meal frequency to extremely high levels to do it; rather they increase the amount of protein consumed per meal. In that larger meals generally take longer to digestion than smaller meals, an athlete increasing the amount of protein at a given meal would simply be expected to take longer to digest that meal. How long might be up to debate based on the data reported above showing increased rate of gastrointestinal transit, but there is no real reason to think that the body can only handle some fixed amount of protein, at least not based on digestion speed alone.

However, an alternate interpretation of the claim is that the body can only optimally utilize a certain amount of protein (i.e. for protein synthesis) with any amount in excess of that being wasted. Up until recently, I was not aware of any research supporting the above idea. However, some recent work has suggested a possible limit in terms of the maximal amount of protein necessary for a maximal protein synthetic response in skeletal muscle.

One recent study fed young and older individuals varying amounts of essential amino acids and examined the response of skeletal muscle protein synthesis. It found a maximal response to 10 g EAAs with no further response at 20 g EAAs (45). Since whole food proteins contain ~40-50% EAAs, this would equate to approximately 20-25 grams of whole protein to maximally stimulate skeletal muscle protein synthesis.

Since the subjects weighed 75 kg, this represents a protein intake of 0.26-0.30 g/kg; a 100kg athlete would be expected to see a maximal response at 25-30 grams of whole protein per meal. An athlete consuming 6 meals per day would be expected to generate a maximal response in terms of skeletal muscle protein synthesis at 150-180 grams of protein per day (1.5-1.8 g/kg of protein).

However, as the subjects in this study were not training, it's unclear if this value can be applied uncritically. As mentioned last chapter, training is known to up regulate a number of important metabolic processes and athletes may have requirements for protein beyond what's needed only to maximize skeletal muscle protein synthesis (46). The topic of protein requirements for athletes is discussed in detail in Chapter 4.

Summary

Protein digestion occurs to a small degree in the stomach with most digestion actually occurring in the small intestine. Specific enzymes "cut" protein into smaller and smaller chains. Generally speaking, only free form amino acids and di- and tri-peptides (two and three amino acid chains) are absorbed; longer chains show insignificant levels of absorption unless some pathological condition exists allowing them to get into the bloodstream unabsorbed.

Following digestion and absorption, the remaining AAs travel to the liver where they undergo extensive metabolism. Over half of the AAs reaching the liver are broken down, a small percentage is used for protein synthesis in the liver and the remainder is released into the bloodstream for use by other tissues.

Animal source proteins generally show very high digestibility, ranging from 95-97% with vegetable proteins showing lower digestion efficiencies of 75-85%; this means that protein requirements will be higher if a large amount of protein is derived from vegetable sources.

Perhaps the most interesting idea in protein nutrition in recent years is that of "slow" and "fast" proteins. Early work suggested simply that whole proteins are generally digested fairly slowly with liquid proteins being digested more quickly than solids. Some protein hydrolysates are digested somewhat faster although the differences are not that large, a few minutes at best. Free form amino acids tend to be digested the most rapidly, although di- and tri-peptides are digested quickly as well.

More recent work has found that there can be other differences between proteins in terms of speed of digestion. The original research examined whey and casein with whey being the prototypical "fast" protein and casein the prototypical "slow" protein. Whey, at least consumed by itself, spikes amino acids levels but there is an equally rapid decrease. This tends to stimulate whole body protein synthesis with a smaller effect on whole body protein breakdown; amino acid oxidation is also stimulated by the rapid increase in blood amino acid levels.

In contrast, casein ingestion leads to a much lower but sustained increase in blood AA levels, this decreases protein breakdown with only a small effect on protein synthesis. Differences between the two proteins are significantly decreased when other nutrients are added to the meal however. It's unclear how the consumption of previous meals will impact on digestion rate as no research appears to have been done on this topic.

The picture gets complicated quickly as different proteins show very different rates of digestion. Additionally, it appears that chronically high food intakes cause adaptations to occur that increase the rate of digestion/absorption of protein (and other nutrients). Protein digestion rates determined in individuals with a habitually low protein intake may underestimate the digestion speed in athletes who consume large amounts of protein each day.

While many have interpreted the "fast" and "slow" protein research as proving the superiority of "fast" proteins, this isn't a necessarily valid conclusion. While fast proteins are probably superior before and during workouts (for practical reasons), there is evidence that slow or fast/slow mixtures may be superior following workout. Whether "fast" or "slow" proteins are superior at other times of the day is currently unknown.

Finally, although there appears to be no limit on the amount of protein that can be digested per meal (larger amounts simply take longer), there is some evidence for a limit to the amount of protein per meal in terms of maximally stimulating protein synthesis. Although this would seem to set a limit on the amount of protein which should be consumed per meal, it's important to note that there are other metabolic processes requiring protein which are unrelated to skeletal muscle protein synthesis per se which might mandate higher protein intakes per meal for optimal adaptation to training.

3

Basic Protein Metabolism

In the last chapter, I discussed protein digestion and absorption up until the point that amino acids (AAs) enter the bloodstream/free pool. Since it will provide a background for several subsequent chapters, I next want to look at some basic concepts of protein metabolism in terms of how those AAs can be used. My primary focus here will be on skeletal muscle since that is the tissue of greatest importance to athletes.

First I want to examine the processes of protein synthesis and breakdown as well as the concept of protein turnover. Next, I want to look at how the simple act of eating a meal impacts on that process; a concept referred to as diurnal cycling is important as it helps to explain why simply eating lots of dietary protein doesn't increase the body's protein stores.

The impact of training will be examined next with the effects of resistance and endurance training on skeletal muscle protein metabolism. Finally, I want to examine the concept of inter-organ "flow" of AAs, how tissues such as skeletal muscle release AA's back into the free pool for use by other tissues. At the end of the chapter, I'll present a model of protein metabolism that ties together the information from this chapter with that from Chapter 2.

Protein turnover: The link between protein synthesis and breakdown

Although the amount of tissue in the body tends to remain fairly constant over time, those tissues are actually undergoing an essentially continuous process of breakdown and resynthesis; the two processes together are generally referred to as tissue turnover.

This holds for protein-based tissues such as plasma proteins and skeletal muscle which undergo a continuous process of breakdown and resynthesis. Fundamentally what occurs in terms of the amount of these tissues present depends on the long-term relationship between protein synthesis and breakdown.

If synthesis exceeds breakdown, there will be an increase in the amount of that protein. If breakdown exceeds synthesis, there will be an overall loss in the amount of that protein. If breakdown equals synthesis, there will be no long-term change in the amount of that protein.

Unless an athlete is specifically trying to lose muscle mass (a rare but not unheard of situation), they either want skeletal muscle protein synthesis to be equal to or greater than protein breakdown. This means either increasing protein synthesis, decreasing protein breakdown, or doing both at the same time.

It's important to note that different tissues turn over at drastically varying rates. Plasma proteins made in the liver may turn over in a matter of hours while skeletal muscle protein may take days to turn over; tissues such as tendons and ligaments may take months or years to turn over completely (1).

The process of protein synthesis requires that AAs be pulled out of the free pool for incorporation into the protein being synthesized; protein breakdown releases AAs back into the free pool. The specific pathways and mechanisms of protein breakdown and synthesis aren't ultimately that important from a practical standpoint. Rather, it is important to note that those pathways are separate and, as you'll see shortly, regulated by different factors.

That is to say, protein synthesis is not simply the reverse process of protein breakdown; nor is protein breakdown simply the reverse process of protein synthesis. Rather, they are distinct physiological pathways that are regulated by different factors in the body. I'd note that, while protein synthesis and breakdown are separate processes mechanistically, they are also interrelated to some degree; under many circumstances, such as growth, increases in protein synthesis are accompanied by increases in protein breakdown as well (1).

Under normal dietary circumstances, an average-sized individual may turn over roughly 300 grams of protein per day; that is a total of 300 grams of protein will be broken down with most of it being resynthesized back into tissue. Larger and smaller individuals will turn over proportionally more or less total protein per day. As I'll discuss in Chapter 4, this doesn't mean that daily protein requirements are 300 grams per day as most of the protein broken down is simply reutilized by the body.

Protein turnover is energetically costly, quite in fact it's been estimated that protein turnover may account for 15-25% of basal metabolic rate (3). At first glance, protein turnover seems a rather wasteful process for the body to undergo, especially since the net result is more or less maintenance of bodily tissue (since most of the protein broken down is resynthesized again). However, protein turnover appears to play a crucially important role in dealing with stressful situations by providing AAs where they are needed.

Reduced protein turnover might compromise the body's ability to rapidly deal with stressful stimuli (1,3,4). For example, the enhanced rate of muscle breakdown seen in burn and trauma patients occurs to provide sufficient AAs (especially glutamine and its precursors) to sustain the immune system (3). Of course, this occurs at the expense of muscle tissue, explaining the muscle wasting seen in such situations.

Protein turnover is mediated by a number of factors. This includes hormonal factors (testosterone, thyroid, insulin, cortisol, GH, glucagon), caloric intake, and AA availability (5). The impact of specific hormones (except for insulin) is beyond the scope of this book. Of course, training has a profound impact on both protein synthesis and breakdown which I'll examine below.

How eating affects protein synthesis and breakdown

On a day-to-day basis, perhaps the single largest impact on skeletal muscle metabolism comes from the simple act of eating a meal; this has the capacity to both stimulate protein synthesis and inhibit protein breakdown.

Although other factors are certainly involved, it turns out that the primary factors affecting protein breakdown and synthesis following a meal are the concentrations of insulin and blood AAs (6) which turn out to play independent but interacting roles (7).

Looking first at protein synthesis, the AA content of a meal plays the major role in terms of promoting protein synthesis with insulin playing a secondary role (8). Quite in fact, assuming sufficient AAs are available, only very small amounts of insulin are required for maximal stimulation of protein synthesis via amino acids.

More directly, insulin also increases AA transport into skeletal muscle and some research has suggested a direct role of insulin on protein synthesis (9). However, this is all contingent on there being sufficient AAs present in the first place; raising insulin levels without raising AA levels (by consuming protein) has little to no impact on protein synthesis (10). In fact, elevating insulin without simultaneously increasing AA availability tends to decrease protein synthesis, due to a decrease in circulating AA concentrations (6).

Recent research has identified the essential amino acids (EAAs) as being key players in stimulating protein synthesis; the inessential AAs appear to have no direct effect on protein synthesis (11) although the full complement of AAs is required for protein synthesis.

As well, the branched chain amino acids (BCAAs) appear to play an extremely direct role in promoting protein synthesis with leucine specifically stimulating protein synthesis (12); quite in fact, AA mixtures lacking BCAAs are ineffective in stimulating protein synthesis (13).

In terms of protein breakdown, the research is much less well developed but meal consumption appears to decrease protein breakdown via a combination of both increased AA availability (especially leucine) along with the increase in insulin (14). Although it plays a fairly minor role in promoting protein synthesis, insulin appears to play a primary role in decreasing protein breakdown (6).

There are two primary points to be taken from the above research. The first is that the combination of increased insulin and blood AA levels has a combined positive impact in terms of net protein gain after a meal.

The second main point is that the mere act of eating tends to result in an overall increase in bodily protein gain. However, it's important to note that, in general, simply eating piles of protein doesn't lead to significant gains in muscle mass or body protein stores in the long term. The reason for this has to do with a process in the body referred to as diurnal cycling.

Diurnal cycling

The reason for the apparent contradiction above has to do with a physiological mechanism called diurnal protein cycling (15). Diurnal cycling refers to the cyclical nature of protein metabolism whereby net protein synthesis during the fed state (when food is being consumed) is matched by net protein breakdown during the fasted state (when food is not being consumed). This seemingly wasteful process is thought to provide meal derived AAs to the body more evenly over a 24 hour period (16).

Although feeding clearly impacts on skeletal muscle protein synthesis and breakdown, many of the proteins synthesized (e.g. in the gut and liver) after a meal are labile proteins; that is, proteins which serve mainly as a temporary storage site for amino acids (16). These labile proteins are broken down between meals (or during the night), leading to essentially no net gain or losses in body protein over a 24- hour period. Diurnal cycling is thought to act as a "buffer" to prevent increases in circulating AAs from occurring after a meal since they are directed into tissue synthesis (17).

Diurnal cycling is sensitive to protein intake. As protein intake goes up, and protein storage during the day increases, there is increased protein breakdown at night. Thus, the more protein an individual eats, the more he or she needs to eat to maintain balance (1, 18). Some researchers have even suggested that the high apparent protein requirements in athletes are being driven by a habitually high protein intake (19). That is, chronically high protein intakes tend to require high protein intakes to avoid losses due to diurnal cycling (and increased AA oxidation in the liver).

By the same token, when protein intake goes down, there is less protein stored during the day, and less broken down at night. However, whenever there is a change from high to low protein intakes, there is a short lag time before diurnal cycling (and other processes such as AA oxidation) "catches up" to the change in protein intake (20).

In any case, the process of diurnal cycling explains why simply consuming massive amounts of protein in and of itself doesn't cause significant muscle gain. The body simply loses more protein between meals in addition to oxidizing larger amounts in the liver. In non-training individuals, this tends to keep the body in a state of maintenance rather than causing an increase in body protein.

When training is added to the equation, the dynamics change and the body uses the extra incoming dietary protein to synthesize new proteins in the body. In essence, proper training "tells" the body to store more protein; combined with sufficient building blocks (calories and protein) this leads to increases either in skeletal muscle mass or in the proteins and enzymes involved in endurance performance.

The effects of training on protein metabolism

Outside of the processes related to normal growth or aging, possibly the single greatest factor influencing skeletal muscle metabolism is training. Although the physiology of resistance and endurance training are significantly different, both impact profoundly on skeletal muscle metabolism.

Resistance training affects both protein synthesis and breakdown with both being increased following training (21). Additionally, protein breakdown is stimulated to a greater degree than protein synthesis; this means that immediately after training the body is in a net catabolic state, breaking down more protein than it is synthesizing.

As I'll discuss in Chapter 8, the provision of nutrients around training is crucial to shifting skeletal muscle back towards a net anabolic state, where more protein is being synthesized than broken down.

Properly performed, resistance training has a net anabolic effect on the body (19), essentially "telling" it to maintain body protein stores at a higher level. Which is simply a complex way of saying that proper resistance training (accompanied of course by sufficient protein and calories) leads to increased levels of muscle mass. As I'll discuss in the next chapter, this is a major part of the increased requirement for protein with strength training: increased dietary protein is required to provide the building blocks for the proteins in skeletal muscle. This is in addition to any other metabolic processes that may be upregulated by training (19).

In contrast, endurance training has a profoundly different impact on skeletal muscle. While resistance training primarily stimulates increases in contractile proteins (muscle tissue), the proteins synthesized following endurance training are primarily enzymes and mitochondrial proteins which enhance energy production during activity.

Additionally, while resistance training is distinctly an anabolic process (increasing the body's protein stores over time), endurance training has far less of an impact in this regards with only a short-lived increase and retention of amino acids in skeletal muscle following training (19); as well, large amounts of endurance training can be significantly catabolic causing muscle size to decrease.

While some of the loss of muscle tissue is simply an adaptive response to endurance training, there is an additional direct effect of long-duration endurance activity on protein metabolism. Specifically, long-duration endurance training increases amino acid oxidation (with the BCAAs and especially leucine being the primary AAs used) to provide a small portion of the energy used (19); this can provide anywhere from 5-10% of the total energy during training. I'll discuss this topic in more detail in Chapter 6. As you'll see in Chapter 8, the typical protein breakdown seen during endurance training can be prevented by the provision of appropriate nutrients around training.

Inter-organ "flow" of amino acids

In a similar way that the liver releases AAs into the bloodstream for use by other tissues, those other tissues can also release AAs back into the bloodstream so that they may be used elsewhere in the body. Skeletal muscle is one of them and has been found to release large amounts of both glutamine and alanine into the bloodstream. Smaller amounts of other AAs are also released but my focus will be on those two.

Early work identified that the amounts of glutamine and alanine released from skeletal muscle actually exceeded the amount that was present in the muscle in the first place; as well, the proportion of those two aminos that were released was far out of proportion to the AA profile of skeletal muscle. Subsequent work determined that skeletal muscle is able to synthesize both glutamine and alanine from other sources.

Skeletal muscle has the capacity to break down several AAs including the branch chain amino acids, asparagine, aspartate and glutamate (16). AA catabolism occurs to a greater degree under such conditions as extreme stress, fasting, or dieting.

Similar to what occurs in the liver, the breakdown of AAs within skeletal muscle produces a carbon skeleton along with ammonia. Since the ammonia can't be made into urea (as would occur in the liver), it is combined with glutamate to produce glutamine. The body actually turns out to synthesize between 20-80 grams of glutamine per day (16). The gut, immune system, liver, kidneys and pancreas are all able to use that glutamine for their own use.

Alanine is produced primarily from the catabolism of the branched-chain AAs and is primarily used by the liver. There, the alanine produced can either be used to form urea or, under conditions of caloric/carbohydrate restriction, be used to produce glucose. As I'll discuss in a later chapter, recent evidence has found that diets high in leucine not only stabilize blood glucose but also spare lean body mass loss when calories are restricted.

Summary

In the first part of Chapter 2, I discussed the process of protein digestion and absorption in the small intestine, metabolism in the liver and subsequent release into the bloodstream. In this chapter, I've examined the basics of protein metabolism, focusing on skeletal muscle, once those AAs have appeared in the bloodstream.

Most tissues in the body are in a constant state of breakdown and resynthesis, a process referred to as protein turnover. Tissues turn over at different rates and it is the final balance between synthesis and breakdown that determines whether or not a given tissue gains or loses protein.

The simple act of eating affects both protein synthesis and breakdown, generally inhibiting protein breakdown and stimulating protein synthesis. Amino acids are the primary regulators of protein synthesis with a direct stimulatory role; they play a secondary role in inhibiting protein breakdown. In contrast, while it has some direct effects on protein synthesis, insulin primarily works by inhibiting protein breakdown. The combination of the two, an increase in blood amino acids along with an increase in insulin concentrations, has the greatest impact on overall net protein gain.

However, simply eating tons of protein generally doesn't generally lead to gains in muscle mass. Rather, protein stored during the day is broken down and released to the body during the night, a process referred to as diurnal cycling. With increasing or decreasing protein intakes, diurnal cycling can increase or decrease after a short adaptation period; over time this ensures that the body simply maintains its protein stores at a constant level.

Increasing the body's protein stores (i.e. skeletal muscle mass) requires a stimulus such as resistance training that, in essence, "tells" it to store more protein. The combination of properly performed resistance training along with sufficient calories and protein leads to an increase in body protein stores; lean body mass is gained.

Endurance training has a much milder effect in this regards, although it stimulates protein synthesis following training, the proteins synthesized are primarily enzymes and mitochondria, both of which are involved in energy production during activity. As well, long-duration endurance training can cause the body to oxidize amino acids, primarily leucine directly for energy. Over time, long-duration endurance training can cause a loss of muscle mass.

In addition to the oxidation of leucine stimulated by endurance training, muscle can also break down several other amino acids. Skeletal muscle has been found to produce both glutamine and alanine in significant amounts from the breakdown of the other amino acids. Glutamine can be used by various tissues in the body or simply broken down to urea and excreted. Leucine is typically used to produce glucose in the liver, especially when calories and carbohydrates are being restricted.

A summary of the various pathways of amino acid metabolism discussed in the previous and this chapter appears in Figure 1 on the following page.

Figure 1: Simplified model of protein metabolism

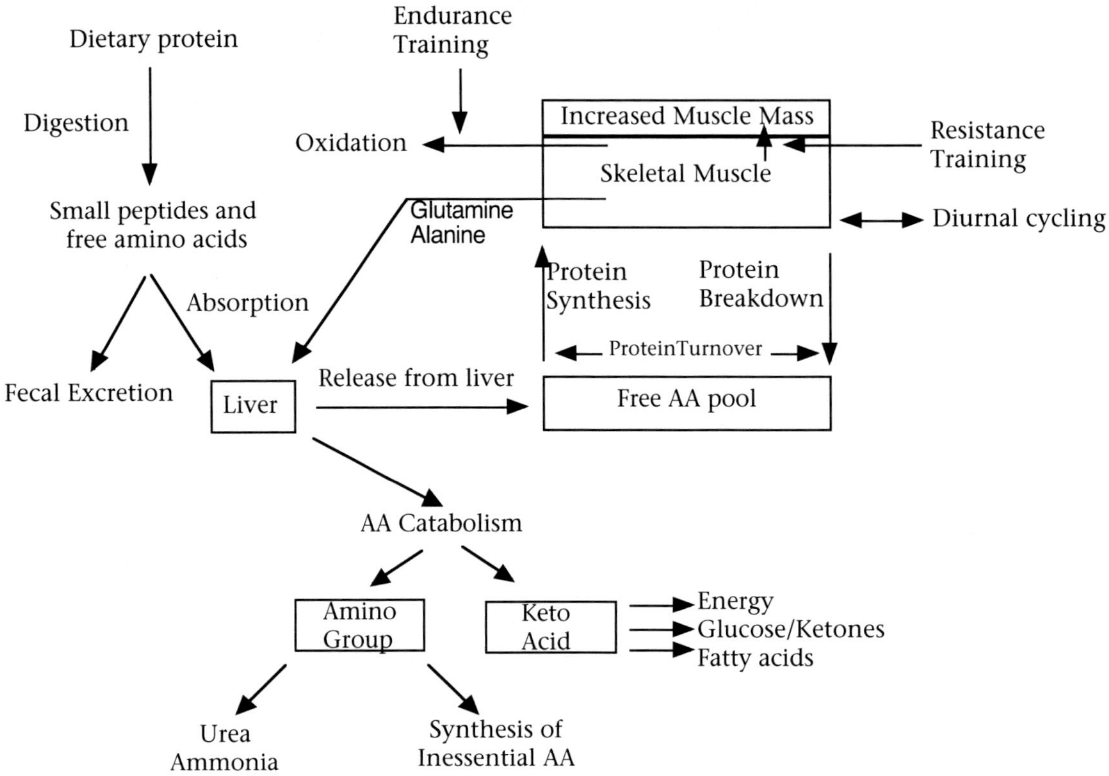

Adapted from Millward, DJ et. al. Physical activity, protein metabolism and protein requirements. Proc Nutr Soc (1994) 53: 223-240 and "Advanced nutrition and human metabolism, 2nd ed." James L. Groff, Sareen S. Gropper, Sara M. Hunt. West Publishing, 1995.

4

Protein Requirements

Debate over human protein requirements has been ongoing since the beginning of nutrition as a science. As research developed over the years, recommendations for optimal intakes for the general population have swung from high to low to high and back again (1). In the area of sports nutrition and requirements for athletes, the argument is just as long-standing with various groups arguing for low, high or in-between amounts for athletes.

As a general rule, mainstream nutritionists have either maintained that athletes don't need any excess protein beyond the requirements for non-athletes, or that athletes already get more than enough in their diet (there is an element of truth to the second statement).

At the other extreme are the athletes themselves, especially bodybuilders and strength athletes, who have long maintained that they need much larger amounts of protein than the general public. Bodybuilding magazines, which usually have close ties to the supplements being advertised, promote this idea as well.

Despite decades of work, there is still argument among researchers over the true human requirements for protein, both for the average non-training person as well as the athlete. Research exists to support both sides of the argument, further contributing to the controversy and disagreement. I don't see a resolution coming about any time soon.

I'm going to cover a number of topics in this chapter. First and foremost I want to address the issue of how protein intake recommendations should actually be set. Then I want to provide a bit of background information on nitrogen balance, which tends to be the primary method used for determining protein requirements.

With that background, I'm going to look at both sides of the debate over protein requirements for athletes, examining both the arguments for high and low intakes. Following that, I'm going to provide what I feel is an optimal compromise for the current argument, along with specific intake recommendations for both endurance and strength/power athletes.

In addition, I want to discuss how protein requirements may change when calories are restricted during a diet. Finally, I'll going to examine some of the data suggesting a possible gender difference in protein requirements for athletes.

How should protein intake recommendations be given?

Before tackling the controversy over protein requirements for athletes, I want to address the issue of how protein intake recommendations should be given. It's quite common to recommend that athletes obtain some percentage of their diet from the macronutrients. For example, an intake of 30% protein, 60% carbohydrates and 10% fat might be recommended.

I disagree with this approach for one simple reason: percentages may or may not have any relevance to the actual nutrient requirements of an individual. For example, consider diets consisting of either 1000 or 4000 calories per day, both containing 30% protein. The first diet would provide 75 grams of protein while the second would provide 300 grams of protein. Although both contain 30% protein there is more than a 3-fold difference in actual protein intake. While many researchers continue to use percentages to recommend protein intake, this is typically within an assumed caloric intake. I don't find this to be a necessarily safe assumption and prefer to use a different method.

It's more appropriate to make recommendations in terms of grams of a given nutrient per kilogram (or pound) of body weight. Thus, a protein recommendation of 2 g/kg (0.9 g/lb) will be the same whether the athlete is eating 1000 calories, 4000 calories, or 10,000 calories per day. The percentages will vary drastically, but are ultimately meaningless: 2 g/kg is 2 g/kg whether that makes up 10%, 50% or 100% of the total caloric intake. Throughout this book, I'll put protein intake recommendations relative to body weight (or in one specific, in absolute terms). As a quick side-note, readers can convert kilograms to pounds by multiplying by 2.2; to convert pounds to kilograms, simply divide by 2.2.

A question that often comes up is whether protein (or carbohydrate or fat) recommendations should be made relative to total body weight or lean body mass (LBM). LBM is determined by subtracting the total amount of body fat from the total weight; everything that is left is considered LBM and this includes skeletal muscle, bone, organs, body water, etc. The distinction between total body weight and LBM becomes increasingly important at higher body fat levels as shown below.

For lean male athletes (i.e. 8-10% body fat), the distinction is essentially a non-issue as total body weight and lean body mass are nearly identical: lean mass represents 90-92% of total body mass in this case.

Consider, however, an athlete who is 100 kg (220 pounds) with 20% body fat. He has 20 kg (44 pounds) body fat and 80 kg (176 pounds) of lean body mass. Let's say he wants to consume 3.0 g/kg (1.4 g/lb) of protein per day. Using his total body mass, his protein intake would be 300 grams per day. Using lean body mass, it would be 240 grams per day, a difference of 60 grams per day.

Furthermore, the differences become larger as total body weight or body fat percentage goes up. A 150 kg (330 pounds) athlete at 30% body fat and 3 g/kg (1.4 g/lb) of protein per day would be consuming 450 grams of protein if he used total body weight but only 315 grams per day if he used lean body mass; a difference of 135 grams protein per day.

Related to this, I should mention that female athletes, as a consequence of having higher average body fat levels than men, would tend to have their protein requirements overestimated to some degree if they simply use total body weight. That is, a male at 10% body fat is the rough equivalent of a female at 19% body fat (the difference has to do with variances in essential body fat levels) and an identical protein recommendation based on total body weight will overestimate the female's true requirements. As discussed below, there is also some reason to believe that females have slightly lower protein requirements than males physiologically. For these reasons, I'll be recommending a slightly lower protein intake for female athletes.

Logically, of course, it makes the most sense that LBM would be the primary determinant of protein requirements, there being little need to provide large amounts of dietary protein to fat cells. However, this has to be weighed against the general difficulty in getting a good estimate of body fat percentage and body fat (necessary to determine true LBM); methods can vary drastically and many are inappropriate for athletes.

For the reasons discussed above and to remain consistent with the research (which always expresses protein recommendations relative to total body weight), I will express protein intake recommendations relative to total body weight throughout this book. I'd simply note that athletes carrying an excess amount of body fat may wish to scale back their total protein intakes slightly to account for the extra fat they are carrying.

Protein turnover and nitrogen balance

In the previous chapter I discussed the topic of protein turnover and mentioned that, in an average person on a typical diet, the body may turnover (breakdown and resynthesize) roughly 300 grams of protein per day (2). As also mentioned in that chapter, this doesn't mean that daily protein requirements are 300 grams because the majority of protein broken down is simply remade into protein tissues.

No process in the body is 100% efficient and protein turnover is no different. After breakdown, some percentage of protein will end up being oxidized (burned) and excreted as urea, creatinine, ammonia or other substances. With normal dietary protein intakes, roughly 4% of the daily protein turned over may be lost from the body (4). That protein has to be replaced by protein consumed in the diet and this small loss of protein due to turnover is a large part of basal protein requirements for humans.

If you recall from Chapter 1, one of the defining aspects of dietary protein is that it contains nitrogen, and researchers measure protein loss by measuring nitrogen loss. Typically nitrogen is lost in the urine, but it is also lost in feces, sweat, hair, skin, fingernails and other tissues (4).

Since it is extremely difficult to measure all sources of nitrogen loss (early studies would sew people into bags so that all excreted nitrogen could be measured and estimates have been developed from those studies), estimates are generally used for feces, skin, hair, sweat and the rest. Generally speaking, in modern studies, only urinary nitrogen excretion is measured, as this represents roughly 80% of the total daily nitrogen loss; other losses are simply estimated from previous work.

Such measurements allow researchers to perform a nitrogen balance study; they are simply comparing the amount of nitrogen going into the body (via dietary protein) to that coming out of the body (urine, feces, etc.). If individuals are consuming exactly as much nitrogen as they are losing, they are said to be in nitrogen balance. If they are consuming more nitrogen than they are losing, they are said to be in positive nitrogen balance and are presumably storing protein in the body. If they are losing more protein than they are consuming, they are said to be in negative nitrogen balance and are losing protein from the body.

As mentioned, protein (amino acids, to be more accurate) is the sole source of nitrogen; therefore a negative nitrogen balance implies that the body is breaking down body protein. However, a simple nitrogen balance study doesn't give any indication of which specific aminos are being lost or where the protein is actually coming from in the body. It could be coming from muscle tissue or from the breakdown of liver proteins for example; under extreme conditions of severe starvation, organ protein (heart, etc.) can be broken down and lost.

As a side note, I should mention that more recent studies have begun to use more accurate methods to track changes in protein metabolism in the body. Amino acids are radioactively labeled and researchers can then examine where specific aminos are going to or coming from in the body along with where they are going. Much of the research that I discuss in Chapter 8 on nutrient timing has been done using this type of methodology.

Nitrogen balance doesn't simply depend on nitrogen intake (and output); other factors such as caloric intake also play a major role in determining nitrogen balance (5). This can make the determination of true human protein requirements difficult as slight deviations in caloric intake from maintenance needs can affect the nitrogen balance measurements and the estimates of protein requirements that are made from them (1, 5).

As calorie intake goes up, so does nitrogen balance (the body retains more protein); as calorie intake goes down, nitrogen balance falls as well (5). As you'll see below, protein requirements go up while dieting and this is due to the body retaining less protein when calories are reduced. Individuals who eat more protein will lose more nitrogen as well, simply from the increased intake of nitrogen (3).

Early research suggested that carbohydrates were better at improving nitrogen balance than fats (6) while more recent research does not support this (5). At maintenance and higher intakes, dietary fat appears to be as effective, if not slightly more effective at improving nitrogen balance than carbohydrates (5, 7).

I'll talk about the problems with nitrogen balance in a moment; for now I want to describe how a typical nitrogen balance study is done. First and foremost, researchers have to determine what are called obligatory nitrogen (or protein) requirements. This is done by first having the individual eat a protein free diet; ideally the individual is in perfect energy balance, consuming just enough calories to support their activity level. Nitrogen excretion is measured and this is taken to represent the obligatory breakdown of body protein; that is the amount of protein that the body loses per day regardless of diet or activity.

Obligatory nitrogen loss has been estimated at roughly 50-60 mg/kg/day so a 100 kg (220 pound) person would lose 5-6 grams of nitrogen per day (8). Since protein is approximately 16% nitrogen, this works out to roughly 33 grams of protein per day (5-6 g divided by 0.16 = 31-37 grams protein). Since dietary proteins aren't digested or utilized with 100% efficiency, a safety factor is added to this; this is how the Recommended Daily Intake (RDI) for protein is established (8). The RDI replaces the old Recommended Dietary Allowance (RDA).

The US RDI for protein is 0.8 g/kg (0.36 grams/lb) of protein per day. Therefore, a 100 kg (220 lb individual) needs 80 grams of protein per day to meet the RDI; a typical 75 kg male (165 pounds) needs 60 grams. The RDI is taken to cover the protein needs of approximately 95% of individuals.

The RDI assumes a few things including high quality protein and sufficient caloric intake (4). As mentioned above, when calories are reduced, the RDI for protein is no longer sufficient; additionally, diets that contain poorly absorbed or low quality proteins require that more total protein be consumed to compensate. I'd mention that the typical American diet, high in animal products, typically provides 2-3 times the RDI in protein to begin with. This is an important point that I'll come back to below.

Finally, I want to mention that, in addition to a requirement for total protein (specifically nitrogen), the body also has a requirement for essential amino acids and these requirements tend to change at different times of life (for example, see references 6 and 7). I discuss the issue of amino acid requirements in more detail in Chapter 6.

Athletes and protein requirements

As noted above, the RDI for protein is determined in non-exercising individuals consuming calories at maintenance. Simply, the RDI was never meant to cover protein requirements for active individuals. To quote from the RDA handbook itself: "No added allowance is made here for the usual stresses encountered in daily living, which can give rise to transient increases in urinary nitrogen output. It is assumed that the subjects of experiments forming the basis for the requirement estimates are usually exposed to the same stresses as the population generally." (Reference 8, pg. 71)

However, intense training is a stress and would be expected to increase protein requirements; multiple studies have suggested that exercise increases protein requirements (11, 12). However, I'd note their use of the word "transient" above; I'll come back to this below.

Based on nitrogen balance data, both endurance and strength/power training have been found to increase protein requirements although they do so for different reasons. As mentioned last chapter, during prolonged aerobic activity, amino acids can be used directly for energy, this is especially true of the branched chain amino acids (BCAAs) with leucine having been the amino acid most studied; protein can provide 5-10% of the total energy yield during prolonged endurance training and this is more pronounced when muscle glycogen is depleted (13,14).

Although amino acids are not used for energy to a significant degree during strength or power training, protein requirements are still increased to cover repair of damaged tissue along with the synthesis of new tissues. In the case of strength/power athletes, this new tissue is typically comprised of contractile proteins; endurance athletes synthesize mitochondrial proteins and enzymes in response to training as well.

As mentioned in Chapter 1, there are a number of additional metabolic processes important to athletes that are likely to be up regulated through athletic training (15). All of these are likely to require increased amounts of dietary protein although what amount is currently unknown. It's possible that the amount of protein required to maintain nitrogen balance (or generate a positive balance) is still insufficient to optimize all aspects of metabolism important to athletes (15).

From a practical/performance point of view, the mechanism for the increase in protein requirements is somewhat less important than the apparent fact that heavy training increases protein requirements. Data reviewed by Dr. Peter Lemon in 1991 suggested protein requirements of 1.2-1.4 g/kg (0.54-0.63 g/lb) for endurance athletes and 1.2-1.7 g/kg (0.54-0.77 g/lb) for strength/power athletes (13). This value is roughly 50-100% higher than the RDI for protein which is 0.8 g/kg (0.36 g/lb).

A more recent review concluded that 1.6 g/kg (0.72 g/lb) probably represents the high end of protein requirements for endurance athletes, and that much only for those involved in top level sport, training 5 or more days per week for an hour or more per day (16). Recreational endurance athletes, training less than that, are likely to need less protein.

As a side note, bodybuilders have long used an intake of 2.2 g/kg (1.0 g/lb) of lean body mass as a rough estimate for daily protein requirements. Others have often recommended higher intakes of 2.5-3.3 g/kg (1.1-1.5 g/lb) and higher recommendations than this are sometimes seen. Empirically, bodybuilders using anabolic steroids feel that they grow better with intakes of 4.4 g/kg (2 g/lb) or higher per day but little research has been done on this topic.

Problems with nitrogen balance

However, not all researchers agree with Dr. Lemon's conclusions. A researcher named DJ Millward, for example, has argued rather strongly against the studies used to reach the values above, citing problems with nitrogen balance as a method (17,18). The basic problem is that the error inherent in nitrogen balance studies tends to make it a less than ideal method, especially for estimating the protein requirements of athletes. Those same errors tend to accumulate, leading to an overestimate of true protein requirements (19).

Millward points out that, if you take the positive nitrogen balance values at face value and extrapolate them out to the amount of muscle that should have been gained by the study subjects, that increased muscle mass simply doesn't show up. For example, based on the positive nitrogen balance in some studies, lean mass gains of 300-500 g per day or 3.5 kg per week (over 7 pounds of muscle) should be seen. The studies didn't find that and the logical conclusion is that the nitrogen balance method is flawed.

Between the estimates needed for losses of nitrogen in skin, hair, sweat, etc., the method simply isn't that accurate. As mentioned above, the errors appear to accumulate and give false results. And while newer methods are allowing researchers to examine changes in protein synthesis and breakdown directly, these haven't allowed them to come up with protein recommendations for athletes yet.

Furthermore, Millward cites earlier research suggesting that regular training improves protein retention; that is, some research suggested that regular training might *decrease* protein requirements by improving the body's utilization of the protein that is ingested (20). That is, there is some indication that regular training enhances retention of dietary protein, which would tend to lower requirements instead of raise them.

However, it's important to note that the exercise intensity used in this research was much lower than typically used by athletes (15); thus its relevance to athletes engaged in high-intensity training is debatable. It's also possible that, rather than reducing protein requirements per se, exercise simply shifts the use of incoming dietary protein so that muscle gets a "greater share" of incoming amino acids (15); this could conceivably leave other important AA using pathways under-supplied if dietary protein intake only meets this minimum amount.

Related to this, even studies showing an initial negative nitrogen balance with training find that the body can re-achieve nitrogen balance in a few weeks (21). It may be that increased protein intakes are only necessary when a new training program is begun, or when a current program is being intensified. As mentioned above, the RDA handbook acknowledges transient increases in protein requirements with activity and this data may support the idea that there is only a short-term increase in protein requirements with training.

Millward has also pointed out that the rate of actual muscle gain is so trivially low, even with steroids, that the amount of protein required to support growth will actually be quite small (22). To put this in practical terms, assume that an athlete were gaining 0.45 kg (1 pound) of muscle per week. Muscle magazine ads notwithstanding, this rate of gain would be excellent for a natural athlete and most will be lucky to gain half of that (0.22 kg or roughly 0.5 pound) on a consistent basis.

That 0.45 kg of muscle contains approximately 100-120 grams of protein (the rest is water, glycogen and support for the contractile elements of the muscle mass). Gained over a 7 day period, 100-120 grams of muscle protein would theoretically require a mere 15-18 grams of extra dietary protein per day.

Of course, dietary protein isn't used with 100% efficiency although data on how efficiently it is used doesn't appear to exist for humans. Let's assume that it takes 3-5 grams of dietary protein to provide the building blocks for each gram of muscle protein gained. The 15-18 grams per day to support a 0.45 kg/week muscle gain then becomes 45-90 grams of protein per day.

Using the RDI above, a 100 kg (220 lb) athlete has a maintenance protein requirement of 0.8 g/kg or 80 grams of protein. If we add the 45-90 grams of protein that might be required to support a pound per week muscle gain, that comes out to 125-170 grams of protein per day. This yields 1.25-1.7 g/kg, essentially identical to Dr. Lemon's 1991 recommendations.

Another researcher, Stuart Phillips has made similar criticisms of Dr. Lemon's conclusions although along a slightly different line of reasoning (23). He addresses the nitrogen balance studies, accepting that they suggest a slightly increased requirement for protein but points out that training also improves protein retention. He mentions research by Lemon showing that, while novice weightlifters may require 1.4-1.5 g/kg (0.63-0.68 g/lb) of protein per day, more experienced lifters may only need 1.0 g/kg (0.45 g/lb) to maintain nitrogen balance. He also mentions research on individuals performing both strength and aerobic training (meaning they have to cover the protein requirements of both activities) indicated potential requirements as high 1.76 g/kg (0.8 g/lb).

Phillips notes that, in surveys of strength athletes, protein intakes of 2.05 g/kg (0.9 g/lb) or more are typically indicated anyhow. One of Phillips' conclusions is that the whole debate is somewhat moot in the first place, since the average strength/power athlete already consumes protein far in excess of what's being recommended. While this is generally true, athletes do often under consume protein. This can occur due to an overemphasis on carbohydrate or when athletes try to avoid dietary fat to such an extreme that they eliminate high protein foods such as meat or dairy.

Endurance athletes are often guilty of overemphasizing carbohydrate to the exclusion of protein. Even there, due to generally high total caloric intakes, many endurance athletes obtain sufficient protein without really trying; surveys find female endurance athletes typically consume 1.2 g/kg with males consuming 1.8 g/kg of protein per day (16). Female athletes, due to their generally lowered caloric requirements (and intakes), in addition to a general drive to eliminate dietary fat, are often found to be consuming insufficient protein on a daily basis.

Finally, Phillips contends that, given adequate calorie intake (and this isn't always the case, mind you), a protein intake of 12-15% will cover all possible protein needs, although he has recognized that increasing protein while dieting to as much as 25% of the total intake is not only necessary but beneficial (24). Putting this mathematically, consider a 100 kg (220 lb) strength/power athlete consuming 4000-5000 calories per day. At 15% protein intake, he will be consuming between 150-187 grams of protein per day or approximately 1.5-1.87 g/kg (0.68-0.85 g/lb) grams per pound body weight, slightly higher than Dr. Lemon's original recommendations but more than sufficient from Phillips' point of view.

Solving the controversy?

It doesn't appear as if the debate over protein requirements is going to be resolved any time soon. This is true both for the general public as well as athletes. As discussed above, both sides of the argument have good research data to bring to the table. Empirically, of course, you can always find an individual athlete who succeeded with either high or low protein intake.

In my view, the best solution to this problem comes from protein researchers Kevin Tipton and Robert Wolfe (25). In that paper, they examine both sides of the argument in some detail, essentially covering the information I described above in terms of data supporting both increased and decreased protein requirements with training. Following that review, they make a number of important points.

The first and perhaps most important is that coaches and athletes are ultimately less interested in scientific arguments and more interested in what will optimize athletic performance. Unfortunately, studies have not generally examined performance per se as an endpoint: they have only addressed issues of protein requirements.

They also point out that the definition of a "requirement" is context specific. What an endurance athlete who may wish to avoid muscle mass gains (yet support performance adaptations) requires will be different than a strength athlete trying to make weight versus a strength/power athlete wanting to gain strength and mass. Answering the question of "How much protein is required?" depends entirely on the context.

Finally, the paper points out that while small changes in either muscle mass or performance may not be statistically significant in research terms, those same small changes may be crucially important in high level sports. While a 1% difference in gains means nothing in a statistical or scientific sense, it can be the difference between first and last place in the real world of high-level sport. However, we might ask how much relevance it has to the typical recreational trainee.

Additionally, they point out that our current measurement technology is simply unable to pick up such small changes in either performance or muscle mass (especially over the typical 10-12 week training study), and that you can't do controlled studies of athletes over the kinds of time frames (a year or more) that would be required to see those changes. That is, you couldn't take two sets of high-level athletes and control their training and diet for the year or more it might take to see a difference with different protein intakes. In keeping with current research, they point out that the timing of protein intake around training may be as important as the total amount consumed, a topic I'll cover in detail in Chapter 8.

Getting into recommendations, the paper points out that a daily protein intake of 2.5-3.0 g/kg (1.1-1.4 g/lb) for strength/power athletes is not harmful, may give small but important performance improvements over the long-term, and will more than cover any needs for protein synthesis (it's conceivable but understudied that anabolic steroids could increase requirements even further). Any excess will simply be oxidized off in the first place; I'd note again that some researchers feel that the end products of AA oxidation may contribute to the overall anabolic drive in the body.

As well, most of the health-risks often attributed to high-protein intake (discussed in Chapter 9) appear to be of little concern (15) and the biggest potential issue with an "excessive" protein intake will be due to insufficient consumption of other nutrients in the diet (i.e. carbohydrates to support optimal training intensity).

Basically, from the standpoint of high performance sport, they argue that it's better to err on the side of too much dietary protein than too little. At this point in time, we may not know for sure that more protein is necessarily better or optimal, but we do know that too little will tend to hurt performance, gains, and recovery.

In a more recent paper, Phillips make a similar argument, pointing out that there may be a difference in the minimal amount of protein required by athletes and the amount that might provide optimum performance or adaptation in terms of maximizing not only muscular adaptation but other important pathways that utilize protein (15). This might explain the current disconnect between research that finds relatively lower protein requirements for athletes and the athlete's feeling that higher protein provides better performance; researchers are looking at determining minimum protein requirements while athletes and coaches are more interested in optimal performance.

For endurance athletes, Tipton and Wolfe point out that previous research has simply found an increase in protein oxidation with intakes above 1.7 g/kg (0.77 g/lb) and contend that there is no reason to recommend protein intakes in excess of 2.0 g/kg (0.9 g/lb) for athletes involved in endurance sports. There is simply no research data to support the idea that intakes above that level confer any advantage. Quite in fact, extremely high intakes of protein (e.g. 4.5 g/kg or 2.2 g/lb) may actually harm endurance performance, probably by limiting adequate carbohydrate intake (26). This is over double the amount recommended in this book.

Thus a protein intake of 1.7-2.0 g/kg (0.7-0.9 g/lb) seems appropriate for endurance athletes. Given their typically high caloric intakes, not only is this level of protein not terribly difficult to obtain, it should leave plenty of room for sufficient carbohydrate and fat to support training. As mentioned above, female athletes have their own set of issues due to generally lowered caloric requirements along with frequent self-restriction of intake; I'll address this in more detail in Chapter 13.

For strength/power athletes, Tipton and Wolfe contend that an intake of 2.5-3.0 g/kg (1.1-1.4 g/lb) protein per day should be more than sufficient to support any and all needs, without causing any negatives. Once again, this assumes adequate calorie and carbohydrate intake. At this point in time, I feel that Tipton and Wolfe make the best case for protein intakes in athletes seeking optimal performance, and theirs is the position I'll be taking throughout this book.

As mentioned above, all of the protein recommendations above assume sufficient energy intake; it's well known that protein requirements go up while dieting and athletes may need to increase their habitual protein intake slightly when they are trying to lose body fat.

Protein needs during dieting

For various reasons, whether it's to make a weight class or simply to reduce unnecessary fat mass, athletes often have to diet. Muscle loss during dieting is always a danger and finding ways to limit or eliminate that loss has been an area of interest for many years. As mentioned above, calorie intake is a known modulator of protein requirements (5). As calories go up, the body retains more protein; as calories go down, the body retains less and more protein is required.

This means that protein intake needs to go up when dieting to cover the additional requirements (27). In overweight, non-training individuals, a protein intake of 1.5 g/kg (0.7 g/lb), roughly twice the RDA, may be required to limit LBM losses while dieting (28).

Ensuring sufficient protein intake while dieting has additional benefits as well. Higher protein intakes tend to increase fullness, may increase caloric expenditure via thermogenesis (29) and help to maintain stable blood glucose levels (30). Higher protein intakes also appear to limit weight regain after a diet (31). Bodybuilders and other athletes have commonly increased dietary protein to high levels when dieting to reduce body fat and this approach is certainly supported by increasing amounts of research along with decades of empirical success.

How much extra protein is required for athletes to maintain mass or performance while dieting is currently unknown. As mentioned above, sedentary individuals may require almost double the RDA, 1.5 g/kg (0.7 g/lb) versus 0.8 g/kg (0.36 g/lb). Bodybuilders have often increased protein by 50-100% over baseline from a traditional intake of 2.2 g/kg (1.0 g/lb) to 3.3-4.4 g/kg (1.5-2.0 g/lb) or even higher while dieting. This is generally coupled with a more extreme reduction in calories than would typically be seen in an athlete trying to lower body fat.

Athletes should generally not create large caloric deficits to lose body fat, as this tends to harm performance. Using a combination of small reductions in food intake along with small increases in activity and accepting a more gradual fat loss tends to maintain performance more effectively. For this reason, and given the already high protein recommendations being made here, there seems little reason to increase protein intake recommendations significantly above the values listed above.

Under conditions of a slight caloric deficit, consuming protein at the high end of the above recommendations or perhaps slightly higher should be more than sufficient. A protein intake of 3.0-3.3 g/kg (1.4-1.5 g/lb) should generally suffice. As mentioned, bodybuilders and other physique-oriented individuals (figure/fitness) will often go to 4.4 g/kg (2.0 g/lb) towards the end of their competition preparation due to the extreme nature of those activities and the extreme level of leanness that is being sought. Without anabolic steroids, it seems unlikely that any athlete would need more than 4.4 g/kg (2.0 g/lb) of protein even under dieting conditions.

Because of the nature of endurance sports, increasing protein slightly while dieting, by perhaps 20% seems a reasonable recommendation (I'm aware of no research on this topic). This would increase protein intake from a habitual level of 1.7-2.0 g/kg (0.7-0.9 g/lb) to 2.0-2.2 g/kg (0.9-1.0 g/lb). Coupled with either a moderate caloric deficit or slight increases in energy expenditure, this should still allow endurance athletes to consume sufficient carbohydrate and fat calories to sustain performance while still limiting muscle or performance losses.

Female protein requirements

As a final issue in this chapter, I'd like to address potential gender differences in protein requirements. There is some data suggesting that female endurance athletes might need less protein compared to males due to the fact that women utilize less protein during exercise than men (32). This effect is probably mediated by differences in hormone levels, especially estrogen (33). Quite in fact, if you give men estrogen, they will burn more fat (and less carbohydrate and protein) during endurance training (34). There is also some evidence that a woman's propensity to use protein during endurance training varies throughout the menstrual cycle (35), which would be consistent with an effect of sex hormones on fuel utilization.

What about strength/power training? There is evidence that estrogen is protective against muscle damage and this may be part of why women tend to experience less muscle soreness on average (39). As well, due to significantly lower testosterone levels (women have, on average, $1/10^{th}$ as much testosterone as men), women do not gain muscle mass as quickly as men. This would tend to suggest lower protein requirements to support strength training; the already low rates of protein synthesis in men will be even lower in females.

In contrast, a recent paper argues that women may use less glycogen and more fat during resistance training than men; this group suggests that females may benefit from slightly higher fat intakes, lowered carbohydrate intakes and suggests identical protein recommendations as for men (37). Related to this, a recent study found that the phase of the menstrual cycle had no impact on skeletal muscle protein or collagen synthesis (38), suggesting that the hormonal influences affecting protein requirements in female endurance athletes may not impact on strength athletes.

Female endurance athletes might conceivably get away with 80% of the intake recommendations for males. This would bring them to roughly 1.36-1.6 g/kg (0.6-0.72 g/lb) of protein per day. I'd note surveys above which show an average protein intake of 1.2 g/kg in these athletes; female endurance athletes may need to pay more attention to their protein nutrition to ensure sufficient intake.

To factor in the differences in body fat percentage (discussed above), along with potential issues of total caloric intake (which I'll discuss further in Chapter 13), I'm going to recommend that female strength/power athletes stay towards the lower end of the recommendations for protein as this should be more than sufficient to support training, protein synthesis, etc. Under most conditions, a protein intake in the range of 2.4-2.5 g/kg (1.1-1.2 g/lb) should be more than sufficient; I'll address exceptions in Chapter 13.

Summary

Human protein requirements have been the subject of intense debate, research, and argument for several centuries with no real consensus having been reached. Research exists on both sides of the argument suggesting that training can either increase protein requirements or even decrease them. Researchers will likely continue to argue about the relative merits or failing of a given individual piece of research for the foreseeable future.

At the end of the day, athletes and coaches are less interested in the academic debates and more interested in what optimizes (or might optimize) performance; little research has examined performance as a specific endpoint.

Given the fact that protein intake at the high end of recommendations is unlikely to have disadvantages and may have benefits that are too small to show up in research (but which may be important to high-level athletes), it seems better to err on the side of slightly too much protein than slightly too little. This assumes that carbohydrate and fat intake aren't shorted in terms of intake due to excessive protein intakes.

Table 1: Recommended protein intakes

Type of athlete	Units	Habitual	Dieting
Male strength/power	g/kg	2.5-3.0	3.0-3.3
	g/lb	1.1-1.4	1.4-1.5
Female strength/power	g/kg	2.4-2.6	2.6-3.0
	g/lb	1.1-1.2	1.2-1.4
Male endurance	g/kg	1.7-2.0	2.0-2.2
	g/lb	0.7-0.9	0.9-1.0
Female endurance	g/kg	1.3-1.6	1.63-1.92
	g/lb	0.6 -0.7	0.75-0.9

My protein recommendations are summarized in Table 1. Readers can either use a calculator to determine their daily protein requirements or simply turn to Appendix 1 which has a chart which has already done it for them.

5

Protein Quality

Similar to the issue of protein requirements for athletes, the topic of protein quality is one of major debate, both in the research world as well as in the realm of sports nutrition and protein supplements.

Arguments are often made that one protein is of higher quality than another, or that protein powders are superior to whole food protein in terms of their quality. Since this is an area of such debate, I want to discuss the issue of protein quality in some detail.

Protein quality refers, in a general sense, to how well or poorly the body will use a given protein. More technically, protein quality refers to how well the essential amino acid (EAA) profile of a protein matches the requirements of the body; the digestibility of the protein and bioavailability of the amino acids (AAs) also play a role (1,2). I'll address specific AA requirements in some detail in the next chapter; here I just want to take a broad look at the methods currently used to measure protein quality.

Methods of measuring protein quality

There are a variety of methods available to measure protein quality. To a great degree, how a protein's quality is rated depends on what method is used. This is part of what allows companies to argue the superiority of one protein over another.

For example, measured by one method, egg protein may be the highest quality protein, but by another method, casein may be superior. This allows two different companies to claim the superior protein, simply by using a different measure. But this raises the question of whether any of the currently used methods of rating protein are valid in the first place, especially when talking about athletes and athletic performance.

Before continuing, I want to make the important point that the quality of a protein is directly related to the physiological needs of the subject being studied (2). The protein that might be optimal for a bodybuilder in a mass phase may not be the same as the protein that is optimal while dieting or for an endurance athlete or for a pure strength and power athlete. As with the issue of protein requirements, which protein is the highest quality is context specific.

Diet and activity affect how AAs are used in the body and dietary proteins serve different roles for different types of athletes (as discussed in the previous chapters). As previously mentioned, long-duration endurance activity tends to oxidize high quantities of the branch-chain amino acids (BCAAs), suggesting that endurance athletes might have a higher BCAA requirement than non-endurance athletes (3). Extra protein for strength/power athletes is typically consumed to support muscle growth and this might require a different AA profile still. I'll discuss the specifics of AA requirements under different conditions in Chapter 6.

In all likelihood, there is no single protein that can be rated as the highest quality protein for all situations. This will become more evident when I examine different whole food proteins and protein powders in Chapters 10 and 11.

The first question to answer is which method of rating proteins is ideal for humans. The short answer is that none of them are ideal, since all make (often incorrect) assumptions, or are based on models that may or may not be applicable to adult human athletes.

The second question is whether the AA requirements of a sedentary individual are the same or different than that of an athlete, whether endurance or strength/power. I'll discuss this in the next chapter.

Although there are numerous different methods to compare proteins, only a few are used frequently enough in popular literature to require discussion. They are chemical score, biological value (BV), net protein utilization (NPU), protein efficiency ratio (PER), and protein digestibility corrected amino acid score (PDCAAS). I want to look at each in terms of what they represent, how the studies are done to determine them, along with examining whether they have any real relevance to an athletic population, especially one consuming large amounts of high quality protein, in the first place.

Chemical score

Chemical score is a method of rating proteins based on its EAA levels (4). To determine chemical score, some protein is picked as a reference and other proteins are rated relative to that reference protein. This is conceptually similar to giving white bread a value of 100 on the glycemic index scale and rating other carbohydrates relative to that.

Classically, egg protein has been used as the reference protein, but this assumes that the amino acid profile of egg is ideal for humans. Recently, other amino acid patterns have been suggested to replace egg based on increasing information about the AA requirements of humans. In 1985, a joint committee on protein requirements suggested an idealized reference AA pattern for human needs (5). This reference pattern has been criticized as being too low in some amino acids (6, 7).

Since chemical score is a relative, and not an absolute scale, it is possible to have values greater than 100. If 5 grams of the reference protein contains 800 mg of a certain amino acid, and 5 grams of the test protein contains 1000 mg of that same amino acid, the second protein would be rated as 125% for that amino acid.

The EAA present in the lowest quantity (relative to what is required) is defined as the first limiting amino acid (a concept I addressed in Chapter 1). The second lowest EAA relative to requirements would be the second limiting amino acid, etc. You could define a third and fourth limiting amino acid if you wanted although, in general, the first amino acid will determine how well or how poorly a given protein is used in the body. In the case where a given dietary protein had a limiting AA below what was required by the body, supplementation of that AA (e.g. consuming additional methionine with a protein which is limited by methionine) or combining proteins with different limiting amino acids should increase the quality of the protein.

Chemical score can also be used to compare a given protein to the amounts required by an individual in a specific situation. This is somewhat more useful in that it takes into account the needs of the individual, assuming they are known. That is, if a given protein provided 100 mg/kg of a certain AA, and an individual actually required 150 mg/kg of that AA, the chemical score would be 0.67 for that amino acid (meaning that the protein in question provided only 67% of the amount required by that person).

While chemical score is useful for rating proteins based on their composition, it has one major drawback: it has little to do with how a food protein will be used in the body since it does not take into account digestibility; rather, it simply compares the AA content of a given protein to some idealized AA pattern in relative terms. As mentioned, this assumes that the optimal AA profile is actually known in the first place. For this reason, chemical score is rarely the only measure of protein quality used to rate a protein.

Biological value (BV) and Net Protein Utilization (NPU)

Biological value (BV) is probably one of the most commonly used measures of a protein's quality. The BV of a protein is given as the amount of nitrogen retained in the body divided by the amount of nitrogen absorbed from that protein (4). Therefore, digestibility of that protein is taken into account. Thus:

BV = (nitrogen retained / nitrogen absorbed) * 100

A BV of 100 would indicate complete utilization of a given dietary protein, in that 100% of the protein absorbed was stored in the body with none lost.

To measure BV, subjects are typically first fed a zero protein diet; this allows researchers to determine how much nitrogen is being lost under basal conditions (the obligatory losses I mentioned in Chapter 2). Then the test protein is fed at varying levels (generally 0.3, 0.4, 0.5, and 0.6 g/kg are fed) and a nitrogen balance study (as described last chapter) is done (8). Studies use different periods on the protein free diet and this is an important consideration in evaluating BV data as the body shows adaptation to habitual protein intake over time. As mentioned last chapter, nitrogen balance is a far from perfect method so the BV data generated can often be flawed.

For example, a study often cited by advertisers to demonstrate the "superiority" of whey protein hydrolysate measured nitrogen balance in rats after three days of starvation (9). In this study, whey protein hydrolysates led to better nitrogen retention and growth than the other proteins studied. What is not mentioned is that starvation affects how well the body will store incoming protein, leading to falsely elevated BV measures. In any case, this study has little bearing to an individual with an habitually high-protein intake.

As discussed in the last chapter, nitrogen balance is an imperfect method in the first place, nor does it give any indication where the protein being stored is going. Nitrogen balance, and hence BV, only give a rough indication of what is happening on a whole-body level (10). Depending on the individual amino acid requirements of a given tissue, it is possible that a protein might optimally support protein synthesis in one organ, such as the liver, while not optimally supporting synthesis in another tissue, such as muscle.

For example, I mentioned back in Chapter 2 that soy and milk may differentially support AA metabolism in the gut versus skeletal muscle respectively. BV alone tells us nothing about this.

Despite what is sometimes claimed, it is impossible to have a BV greater than 100. Some early advertisements claimed that whey protein has a BV of 157, but this would imply that 1.57 grams of nitrogen were stored for every 1 gram of nitrogen consumed. Since it's clearly impossible to store more nitrogen than is consumed, a claimed BV greater than 100 is also an impossibility.

As it is based on a nitrogen balance study, one aspect of measuring BV that can cause problems in interpretation of results is that the BV of a protein is affected by a number of factors. As discussed in the last chapter, the first of these is caloric intake (10). A very high caloric intake will improve nitrogen balance at any given protein intake and vice versa. This means that an individual consuming a lot of calories (e.g. a strength athlete on a mass-gaining diet) will show improved nitrogen retention and "apparent" BV will go up. By the same token, if calories are decreased (e.g. during a diet), "apparent" BV will go down. A secondary factor that affects BV is activity. Exercise, especially weight training, increases nitrogen retention which will give a protein a higher apparent BV.

A third factor, and one that is typically ignored in popular literature is that the BV of a protein is related to the amount of protein given (8). As mentioned above, BV is measured at levels below the maintenance level. As protein intake goes up, the BV of that protein goes down. For example, milk protein shows a BV near 100 at intakes of 0.2 g/kg (0.44 g/lb). As protein intake increases to 0.5 g/kg (1.1 g/lb) BV drops to 70 or so (8), with higher intakes, BV would be expected to go down even further.

And while some nutritionists have used this fact to try and argue for low intakes of protein to maximize BV, this is an incorrect conclusion. Even with a lower "effective" BV, a greater amount of nitrogen will still be stored in the body at the higher intake. That is, seventy percent (BV of 70) of 0.5 g/kg is 0.35 g/kg, which still means higher nitrogen retention than one hundred percent (BV of 100) of 0.2 g/kg which is only 0.2 g/kg retained.

To quote from Pellett and Young, ".... protein is utilized more effectively at suboptimal levels than at levels in the near-maintenance range of intake. Accordingly biological measures of protein quality conducted at suboptimal levels in either experimental animals or human subjects may overestimate protein value at maintenance levels." (8) Therefore, while BV may be important for rating proteins where intake is below requirements, BV has little bearing with high intakes of dietary protein.

Given that BV is measured at very low protein intakes, it's difficult to see how BV would be terribly relevant to athletes consuming the amounts of protein recommended in this book. A possible exception might be dieting, as the reduced caloric intake would mandate that higher quality proteins be consumed. Even there, as athletes typically increase protein while dieting, far above the levels fed to determine BV, it seems unlikely that BV would be a relevant measure.

A protein scoring method that is very similar to BV is Net Protein Utilization (NPU). Both measure the amount of nitrogen retained in the body with one critical distinction: NPU compares the amount of nitrogen retained to the amount ingested whereas BV compares the amount of nitrogen retained to the amount that is actually absorbed from the gut (1). NPU is simply BV without the correction for digestion or fecal excretion. Since only the amount of protein absorbed from the gut is particularly relevant to fulfilling protein needs, BV is a slightly superior method of measuring protein quality than NPU although neither is likely to have much relevance to athletes consuming large amounts of protein.

Protein efficiency ratio (PER)

PER represents the amount of total weight gained in grams relative to the amount of protein consumed in grams (3). For example, a PER of 2.5 would mean that 2.5 grams of total weight was gained for every gram of protein ingested. A PER greater than 2.7 is considered a high quality protein source (1). Since it is impossible to measure weight gain in grams in humans, PER is generally measured in young, growing animals placed on a diet which is 10% protein by weight. The method by which PER is determined basically excludes it as an appropriate method for rating proteins for adult human athletes on high protein intakes.

And while the Food and Drug Administration has suggested the use of PER with casein as a reference model for labeling protein foods, the use of PER to estimate human protein quality has been criticized and rightfully so (11, 12).

While the use of PER to rate proteins for humans is debatable, it's interesting to note that a recent animal study found that combinations of animal (30% of total) and plant based proteins (70% of total) yielded a higher PER value than the animal or vegetable proteins eaten alone (13). This may have to do with the proteins "combining" to decrease the impact of the limiting AA.

Athletes who wish to decrease their intake of animal-based proteins may be able to achieve higher protein quality levels (at least as measured by PER) with a combination of animal and plant based proteins than someone eating only animal based proteins. Beyond that, I don't feel that PER has much relevance to human nutrition since it is based on studies of young growing animals on low-protein intakes in the first place.

Protein digestibility corrected amino acid score (PDCAAS)

PDCAAS is the newest method of protein quality to be developed. It has been suggested as the ideal scale to rate proteins for their ability to meet human requirements (14). Similar to chemical score, PDCAAS rates protein foods relative to a given reference protein. In this case, the AA profile used is the one determined to be ideal for children two to five years old; this is taken to be representative of the optimal AA profile for adults (15). This raises the immediate question of how much relevance this AA profile has to different types of athletes. I'll address this topic in more detail in the following chapter.

PDCAAS goes beyond chemical score, however, by factoring in the digestibility of a given protein, giving the AA profile more relevance to human diets. Interestingly, using the PDCAAS method, along with the proposed AA reference pattern, proteins which were previously rated as poor quality by other methods, such as soy, have obtained higher quality ratings (15). This is more in line with studies showing that certain purified soy proteins, such as Supro, can maintain adults in nitrogen balance (2,15).

Once again, the use of PDCAAS to rate proteins for adult athletes is debatable since different types of athletes could conceivably require different optimal AA patterns as discussed in the next chapter. Another limitation of PDCAAS has to do with the fact that 1.00 is taken as the highest value that a protein can score; any values above that are simply rounded off to 1.00. This might cause the PDCAAS to under-represent the true protein quality of some proteins since values above 1.00 are rounded down.

Summary

Although a variety of methods of measuring protein quality have been proposed, I feel that none are perfect in rating proteins for human use. While some methods of rating protein are based on how well (or poorly) an animal grows (or the nitrogen balance which is attained), these methods provide no information on specific amino acid requirements or protein synthesis in a given tissue. Rather, only data regarding protein utilization by the whole body are obtained.

Another strategy to rate proteins is to compare the AA profile in food protein to some reference protein. Previously, food proteins such as egg or milk were used as a reference but there has been a recent move toward the use of an idealized reference pattern of AAs to rate proteins. This assumes that the true requirements for a given AA are known to develop an appropriate reference pattern.

Ultimately, all the methods of rating protein quality described above are insufficient for rating proteins for athletes. They are used primarily to determine minimum requirements to either support optimal growth in children (which differs from how protein is used to support training adaptations in athletes) or maintenance in adults. None were developed for use in sports nutrition in the first place

Although I don't consider any of the methods discussed in this chapter to be terribly relevant to athletes, Table 1 below presents a summary of the protein quality rankings of a number of different common food proteins. Since PDCAAS is currently considered the "best" method of rating proteins, I've ranked the proteins by their PDCAAS value from highest to lowest. I'd mainly point out that, with the exception of soybeans, animal source proteins rank considerably higher on the PDCAAS scale compared to vegetarian proteins.

Table 1: Protein quality rankings of some common proteins

Protein	PER	BV	NPU	PDCAAS
Milk	2.5	91	82	1.00
Whey protein	3.2	100	92	1.00
Whole egg	3.9	100	94	1.00
Soybeans	2.2	74	61	1.00
Casein	2.5	77	76	1.00
Beef	2.9	80	73	0.92
Chick peas				0.69
Kidney beans				0.68
Peas				0.67
Pork sausage				0.63
Pinto beans				0.61
Rolled oats				0.57
Black beans				0.53
Lentils				0.53
Peanuts	1.8			0.52
Whole wheat				0.40
Wheat gluten	0.8	64	67	0.25

Adapted from: US Dairy Export Council, Reference Manual for US Whey Products 2nd Edition, 1999.

In the next chapter, I'll further examine the issue of protein quality by looking at AA requirements. Various dietary proteins will be examined within this context.

6

Amino Acid Requirements

In the last chapter, I discussed the issue of protein quality, looking at all of the currently used methods of rating proteins. Recapping, I feel that none of the currently available methods for rating proteins is ideal for, or even particularly relevant to, athletes.

In this chapter, I want to address the topic of individual amino acid (AA) requirements under different conditions as this is arguably more important in terms of determining if any one protein is going to be superior to another for different aspects of athletic performance.

In examining the issue of individual AA requirements, there are two slightly separate issues that I want to address. The first concerns that of maintenance requirements; that is, the AA profile needed to maintain current body protein stores.

Most protein research deals with this topic, albeit usually for groups that are receiving only small amounts of a single low-quality protein; this is frequently accompanied with an insufficient caloric intake in the first place. What I've described is the situation found in many third world countries. Under those conditions, even small changes in protein quality can have profound differences in health and functioning; researchers are always looking for ways to maximize the overall health of such individuals with the least total cost. In many cases, simply adding a small amount of a single AA (i.e. the first limiting AA) can drastically improve protein quality for these groups.

However, this research is unlikely to be relevant to an athlete consuming a large amount of high quality protein from varied sources; the varied AA patterns coupled with the sheer quantity of protein being consumed makes all of that data fairly irrelevant. Still, examining how different high quality proteins relate to the possible maintenance AA requirements of human adults provides a starting point to address the second issue.

That issue is the impact that different kinds of training have on AA requirements and how that might impact upon determining the ideal protein to support specific types of training in terms of adaptations, recovery or growth. As discussed in Chapter 4, it is fairly well established that training increases the total amount of protein required but does the profile of the protein required vary for different situations?

Will a strength/power athlete trying to gain muscle mass require a different AA profile than an endurance athlete who is trying to maintain muscle mass and offset the protein used during their training? Does dieting while training to maintain lean body mass require a different optimal AA profile than that required for maintenance? Or can athletes simply consume more total protein and not worry about the AA profile of that protein?

To address these issues, I'm going to examine the changes in AA metabolism that occur during exercise. While numerous studies have examined the effects of endurance training and other forms of stress such as surgical trauma on AA metabolism, there is considerably less data on the requirements of strength/power athletes or bodybuilders in terms of AA metabolism. Research into dieting is somewhat limited as well.

Determining maintenance AA requirements

The determination of human AA requirements has been the subject of considerable research for several decades. And while improvements have been made in the determination of those requirements, there are numerous methodological problems associated with determining true AA requirements in human subjects (1).

Numerous assumptions regarding the model used, the radioactive tracer used, etc. have to be made to do research in this area. As well, since the body can show adaptations to both low and high protein intakes, it tends to be very difficult to determine true human AA requirements. To quote one researcher in the field: "In my view, definition of adult [essential AA] requirements for protein quality scoring is not currently possible or likely to be useful." (2)

Compared to glucose and fat metabolism, where researchers only have to follow the metabolism of one nutrient, AA metabolism is considerably more complex. Ideally, each of the 20 AAs, many of which can be metabolized and inter-converted, would have to be traced into different tissues, each of which shows drastically different AA utilization. In practice, researchers typically pick one or two AAs to follow and assume that this is representative of metabolism of all AAs in that tissue. However, the choice of tracer used determines the results obtained, making research in this area methodologically difficult (3).

What this means is that current technology (based on the use of radioactive tracer) can only provide small bits of information regarding the topic of AA requirements. In this vein, future advances in research technology may leave much of the information in this chapter incorrect.

Growing vs. non-growing subjects and protein requirements

As discussed in Chapter 3, although there is always a constant breakdown and synthesis of tissues going on, the average adult human or animal does not grow appreciably and bodily tissues are more or less maintained over time. This means that most of the incoming protein is being used for maintenance of current tissue, not for the synthesis of new tissues. Even in children, who are growing fairly rapidly, the growth component of total protein requirements is 15% or less with the remainder of protein requirements being used for maintenance (4).

This difference in metabolism between children and adults appears when one examines total protein requirements, and especially essential AA (EAA) requirements, in infants versus adults. Thus, it seems appropriate to compare protein and EAA requirements in a variety of subgroups to see if any patterns develop.

Table 1 below compares protein and EAA requirements in a variety of age groups based on current Food and Agriculture Organization (FAO) and World Health Organization (WHO) recommendations (5); current recommendations for adults are shown under Adult (a). It should be noted that the FAO/WHO recommendations for AAs has been criticized, and recent research shows that the adult values for EAA requirements may be as much as three times greater than predicted (6-10); this is taken into account under Adult (b) in Table 1. Additionally, the protein requirements suggested for athletes by Tipton and Wolfe are presented for comparison purposes (11).

Table 1: Protein and essential AA requirements

Age group	Protein requirement (g/kg)	EAAs (% of total)
Infant	1.8	43
Preschool child (2 years)	1.2	32
School child	1.0	22
Adult (1)	0.6	11
Adult (2)	0.6	33
Strength athletes (3)	2.5-3.0	Undetermined
Endurance athletes (3)	1.7-2.0	Undetermined

Source for data:
1. FAO/WHO/UNU. Energy and protein requirements. Report of a joint FAO/WHO/UNU expert consultation. WHO Tech Report Ser 1985; 724.
2. Young, V. 1987 McCollum lecture. Kinetics of human amino acid metabolism: nutritional implications and some lessons. Am J Clin Nutr (1987) 46: 709-725.
3. Tipton KD and RR Wolfe. Protein and amino acids for athletes. J Sports Sci. (2004) 22 (1):65-79.

As it likely represents the highest EAA requirement for humans, the FAO/WHO actually uses the EAA requirements of 2-year old children as the reference pattern for adults (12). Any dietary protein that meets or exceeds the EAA requirements for two year olds should readily meet the maintenance EAA requirements for adults as well.

Table 2 on the next page compares the EAA profile of a number of common dietary proteins to the reference pattern used by the FAO/WHO. All of the values listed below represent the mg of a given amino acid found in one gram of dietary protein. So, for example, egg contains 22 mg of histidine per gram of protein.

Table 2: Comparison of AA requirements to amounts found in dietary proteins

AA	Reference Pattern (1)	Human Milk	Egg	Cow's Milk	Beef	Whey Hydrolysate (2)	Soy Isolate (3)
Histidine	19	26	22	27	34	16	NR
Isoleucine	28	46	54	47	48	54	49
Leucine	66	93	86	95	83	89	82
Valine	35	55	66	64	50	82	48
Lysine	58	66	70	78	89	88	64
Methionine+ Cysteine	25	42	57	33	40	32	26
Tyrosine+ Phenylalanine	63	72	93	102	80	65	92
Threonine	34	43	47	44	46	65	38
Tryptophan	11	17	17	14	12	22	14
Total w/o Histidine	320	434	490	477	445	417	403

NR = not reported

Source for data:
1. National Research Council. Recommended Dietary Allowances, 10th ed. National Academy Press, 1989.
2. Boza, JJ et. al. Nutritional value and antigenicity of two milk protein hydrosylates in rats and guinea pigs. J Nutr (1994) 124: 1978-1986. Note that the AA profile may vary slightly depending on how the whey is produced
3. Young, VR. Soy protein in relation to human protein and amino acid nutrition. J Am Diet Assoc (1991) 91: 828-835.

Clearly, looking at Table 2, it's clear that all of the dietary proteins examined match or exceed the EAA requirements listed in the reference pattern. This indicates that all should be more than sufficient in terms of meeting maintenance human AA requirements. As well, as mentioned last chapter in the section on the PDCAAs, and as clearly shown above, soy protein isolate is sufficient to meet adult maintenance EAA requirements.

In any case, based on Table 2 above, there is little reason to believe that one high-quality protein will show any benefit over another high-quality protein in terms of maintaining body protein stores as they all contain EAAs in excess of what is required.

Before moving on to the next topic, there is a claim sometimes made regarding proteins that Table 2 can address. It is sometimes argued that a given protein is superior to another because it has higher quantities of a given EAA, or some combination of EAAs.

While this might be true if we were comparing a high-quality protein to a lower-quality protein (defined here as one that did not meet or exceed EAA requirements), Table 2 shows that not only do all high quality proteins more than meet estimated EAA requirements, but the differences between proteins are fairly minimal in the first place. So whether whey has a higher proportion of EAAs than milk would seem to be rather irrelevant, since both are far in excess of what is required, at least in terms of overall maintenance protein needs. Readers should keep in mind that, as discussed previously, AAs in excess of needs are simply oxidized off in the first place. I would mention that many listings for whey show that the BCAA content is higher, averaging 25% of the total AA content as compared to only 15-20% in other dietary proteins.

Which, once again, isn't to say that one protein might not be better for a given situation than another based on some other aspect of the protein; speed of digestion, the presence (or absence) of other nutrients and cost are all aspects that go into determining what protein might be best. But based simply on EAA content and profile, and looking only at maintenance needs, no one high-quality protein would seem to be superior to any other; they all more than meet human requirements and the differences between them simply aren't that large.

But ultimately, this brings us back to the question raised in the introduction to this chapter: do athletes require a different AA profile than sedentary individuals? Does this differ for strength/power athletes versus endurance athletes? Is there any impact of dieting on AA requirements or what protein might optimally spare protein loss?

Exercise and AA requirements: Introduction

As discussed in detail in Chapter 4, both strength/power and endurance training increase overall protein requirements, although they probably do so for different reasons. Research has clearly shown that mixed-muscle protein synthesis (a measure of all of the proteins synthesized within muscle) increases following both types of training, but the specific types of proteins synthesized are likely to be different (11).

As mentioned previously, with endurance training there is an increase in AA oxidation for energy during activity itself. And while endurance training generally doesn't stimulate increases in the contractile proteins in muscle, there is an increase in mitochondrial protein as well as synthesis of enzymes involved in energy production within the muscle.

With resistance training, increased protein requirements are most likely needed to cover both the breakdown of old proteins as well as the synthesis of new contractile proteins; resistance training has been shown to increase both protein synthesis and breakdown supporting this contention.

Again, I'd mention that training is likely to increase the AA requirements of a number of additional metabolic pathways important to athletes; how much or what specific AAs might be needed for these processes is currently unknown.

With that background, I want to look at how the different types of training as well as dieting might potentially impact on AA requirements and affect what protein might be "ideal" under different situations.

Endurance exercise and AA requirements

As I mentioned in Chapter 3, skeletal muscle can directly oxidize a number of amino acids: the branch chain amino acids, asparagine, aspartate and glutamate (16). During exercise, leucine oxidation has been shown to go up and this effect increases when muscle glycogen is depleted. However, the primary AAs released from muscle are glutamine and alanine. As discussed previously, glutamine is produced to help buffer the production of ammonia while alanine is produced from the breakdown of BCAAs in skeletal muscle.

Related to this, a recent study found that BCAA supplementation prior to exercise resulted in increased muscular uptake of BCAAs; the muscle also released more glutamine and alanine (17). Two observations come out of this study. The first is that muscle is clearly using BCAAs to produce glutamine and alanine. Second, and perhaps more importantly, provision of supplemental BCAAs (or perhaps a whole protein high in BCAAs such as whey) during endurance exercise might spare the breakdown of muscle tissue to provide those aminos.

That is, if the body has a dietary source of BCAAs during endurance training, this might spare the need to break down BCAAs stored in the body (in muscle tissue). As I'll discuss in some detail in Chapter 8, accumulating research suggests that the consumption of small amounts of protein during endurance training limits muscle protein break down.

Related to the above, plasma glutamine levels often become depressed with endurance training and, due to glutamine's primary role in immune function, this can cause immune problems. As I'll discuss in Chapter 12, both glutamine and BCAAs (which acts to "protect" the body's glutamine stores) has shown promise in preventing immune system depression with endurance training.

The above data suggests that endurance athletes might need a larger intake of BCAAs or glutamine as part of their increased protein needs from training. As mentioned above, I'll discuss this topic in some detail in Chapter 12 on supplements.

It's currently unclear whether supplemental BCAAs are of use to endurance athletes given sufficient protein intake in the first place; most research examining supplementation has used subjects consuming significantly less protein than recommended in this book. Given the generally high BCAA content of all high quality proteins (15-25% of the total AA content), it seems likely that simply consuming sufficient amounts of protein in the first place should cover most of the increased BCAA needs of endurance athletes and their training.

However, even with sufficient dietary protein (1.6 g/kg), one study did find that additional BCAAs (12 grams/day for 2 weeks followed by 20 grams immediately before and after training) decreased indicators of muscle damage following 2 hours of bicycling. Endurance athletes involved in particularly intensive or extensive training might consider consuming extra BCAAs in addition to ensuring protein intake at the recommended levels. Increasing total protein intake to the highest levels recommended in this book (2.0 g/kg for males) would also increase BCAA intake.

The glutamine content of most dietary proteins is not extremely high, averaging 4-8% (so in 100 grams of protein, there might be 4-8 grams of glutamine). Glutamine is found in fairly high concentrations in milk, meat, soy, and wheat protein (of all things); whey is somewhat lower in glutamine and eggs are a poor source. Given this fact, endurance athletes who wanted to use glutamine would be better off supplementing.

As mentioned above, and discussed further in Chapter 12, a high BCAA intake spares glutamine in the body and this may be a better overall strategy to protect immune function. Additionally, consuming carbohydrate during training limits many of the immune system problems that can occur. Guidelines for nutrient timing around training are given in Chapter 8.

At this point, outside of the issues raised above in terms of BCAAs and glutamine, there is no research to indicate that any specific dietary protein or AA profile will provide more of a benefit than another in terms of endurance performance or adaptations. I am unaware of any research suggesting what an ideal AA profile to support mitochondrial synthesis or enzymes might be. The high BCAA content of whey (in addition to other benefits such as a high cysteine content) might give it some benefit to endurance athletes; I'll talk more about this in later chapters.

Strength/power training and AA requirements

In contrast to endurance training, protein makes an essentially nonexistent contribution to energy production during weight training or most strength/power sports events. Rather, increased requirement for protein occurs to cover the breakdown of tissue protein and synthesis of new contractile proteins.

As discussed in Chapter 4, the amount of protein needed to cover daily synthesis of muscle in non-drug using athletes is extremely small; even steroid users don't appear to synthesize muscle protein at a rate that demands the types of protein intakes that athletes believe is necessary (18). However, this doesn't explain the empirical observation made by athletes that they do grow better with higher protein intakes, or the apparently higher requirements suggested by nitrogen balance studies (11,19). I'd refer readers back to Chapter 4 for a more detailed discussion of both topics.

However, synthesizing new proteins is not the only requirement for dietary protein during strength training and looking only at protein synthetic rates would seem to be ignoring several important pathways of protein metabolism. In addition to the other pathways of AA metabolism important to athletes mentioned back in Chapter 1 (20), at least part of the increased protein requirements is going to repair the breakdown of tissue that occurs during training.

Unfortunately, the exact amount of tissue broken down has not yet been quantified and how much protein might be needed to cover that demand is also unknown. Nor has researched examined whether specific AAs are broken down or degraded during resistance training; determination of what might be an optimal AA pattern to support resistance training is therefore impossible. However, we might be able to make some educated guesses regarding specific AAs based on what is known about AA metabolism during resistance training.

Although weight training, in general, does not use protein for energy, glycogen depletion is known to activate the enzyme involved in oxidizing BCAAs (21). Therefore, it seems plausible that the glycogen depletion seen with weight training might increase BCAA oxidation. This would tend to be more true for bodybuilding (medium repetition) training than low repetition power training with recent work suggesting more glycogen depletion during resistance training than was previously thought; depending on the volume performed, muscle glycogen can be depleted by 30-40% from baseline following a single bout of training (22). This could conceivably increase BCAA degradation, especially if dietary carbohydrates are restricted and glycogen levels are not replenished.

Given the higher protein recommendations for strength/power sports coupled with the high BCAA content of all high-quality proteins, it seems difficult to see how extra BCAAs would have much of an impact. That is, consider a 100 kg (220 pound) strength power athlete consuming the high end of this book's protein recommendations or 3.0 g/kg; that already provides 300 grams of high quality protein per day. The typical protein contains 15-20% BCAAs with some forms of whey containing up to 25%. Without any supplementation, that amount of protein will provide 45-75 grams per day of BCAAs (more if whey makes up a significant amount of the athlete's total intake).

It would likely take massive doses of extra BCAAs to exert an effect past that. I should note that, empirically, some athletes and strength coaches swear by high dose (typically 0.2-0.4 g/kg around training) BCAAs to enhance growth, recovery and adaptation. I'll address this issue somewhat in Chapter 8 when I look at optimal nutrition around training; BCAAs are discussed in detail in Chapter 12.

I'd note that ensuring optimal muscle glycogen stores and providing carbohydrate during a workout (such as with the consumption of a dilute carb drink) should prevent any oxidation of BCAAs during exercise (21). The provision of protein before and/or during workout may also limit protein breakdown; this is discussed in more detail in Chapter 8.

With certain types of strength/power training there can be large increases in lactic acid; due to its role in modulating acid-base balance (23), this could potentially affect glutamine status. An increased need for glutamine under these conditions could conceivably deplete body stores or cause skeletal muscle to oxidize BCAAs, increasing the needs of either.

I'm not aware of any direct research on the topic, although some studies have examined plasma glutamine levels in different types of athletes. Compared to endurance athletes, powerlifters show lower plasma levels of glutamine, suggesting a fundamental difference in glutamine metabolism compared to endurance athletes (24). A recent study examined the impact of heavy eccentric training on glutamine levels and found no impact (25), suggesting that strength/power training has no real impact on glutamine stores. At this point, the importance of glutamine for strength/power athletes is highly debatable and most direct studies have shown no positive impact of supplemental glutamine. Glutamine is discussed in more detail in Chapter 12.

As mentioned above, in addition to supporting the synthesis of new contractile tissue, increases in dietary protein requirements are also likely to occur to cover repair of damaged or broken down tissue as well. Outside of issues related to BCAA/glutamine metabolism during training, this raises the question of what dietary protein or AA profile might optimally support these processes.

Going back to the concept of maintenance needs, some protein researchers have suggested that human AA requirements be based on the mixed tissue AA profile in the body (4). That is, since maintenance of existing tissue is the goal of dietary protein intake in most individuals, it makes sense that the AA profile needed would match that of the tissues in the body.

This idea might be extended to suggest that the optimal AA profile for the additional protein needed by strength athletes to support growth is the same as that found in muscle in the first place. In this regard, one researcher has argued that whey will be superior for muscle growth for this reason, asserting that it contains an AA profile almost identical to that of human muscle (26).

Arguably the protein with the closest AA profile to human muscle is animal muscle (i.e. meat) and this type of argument has been used as "proof" that meat protein will build the most muscle. Additionally, the occasional protein powder has been developed based on the AA profile of human muscle.

However, I find these types of arguments unconvincing for the reasons discussed in Chapter 2 on protein digestion; the AA pattern that appears in the bloodstream has only a minor relationship to the AA profile of the dietary protein eaten. The liver acts essentially as a gate to ensure that the AAs which are required by the body are released into the bloodstream while any that aren't needed are simply disposed of via oxidation. Even if a protein with the absolutely identical AA profile to skeletal muscle was consumed, this in no way guarantees that AAs in that proportion will appear in the bloodstream in the first place.

Which basically points towards the conclusion I drew in Chapter 5 when I discussed protein quality: my feeling is that, as long as athletes obtain sufficient amounts of both protein and EAAs by consuming the recommended amounts of varied high quality protein, there is little reason to believe that any one protein will have a greater impact on growth than any other based on the EAA profile alone. Put more simply, quantity would seem to be more important than quality at the intake levels recommended in this book.

I'd mention that this doesn't automatically hold around training specifically for reasons I'll discuss in Chapter 8. But on a meal-to-meal basis, as long as sufficient amounts of high-quality protein are being consumed, it seems unlikely that one high-quality protein will show any difference in terms of mass gains or adaptation compared to any other.

What little direct research has compared proteins in strength trainers has borne this idea out. Soy and whey had equivalent effects on muscle mass gains over 6 weeks in one study (27) although another found that milk protein showed a slight benefit over soy in terms of mass gains (28). A recent study found that milk generated better mass gains than soy over 12 weeks when combined with resistance training (29).

Both casein and whey stimulated similar gains in protein synthesis following training although that study didn't examine the long-term effects on mass gains (30). In an unpublished abstract, whey, soy, casein or maltodextrin (all given at 0.7 g/kg/day) all had the same impact on muscle size and strength (31). As mentioned previously, some researchers feel that a casein/whey mix (or milk) might give the best benefits overall (32); emerging research discussed in Chapter 8 supports this idea.

Rather than becoming overly concerned with the overall AA profile, I feel that athletes of all types should primarily focus on getting a sufficient quantity of high quality protein from mixed sources. This is especially true when calories are at maintenance or above, where issues of quality or specific AA profile will become that much less important. Dieting is a separate issue, discussed below, and there is some evidence that different proteins might be optimal around training; this is discussed in Chapter 8.

Paying three times as much for a protein powder that, at best, might give a 1% theoretical advantage over another seems misguided to me. For an athlete consuming protein at the intakes recommended in this book, 2.5-3.0 g/kg (1.1-1.4 g/lb) from high quality mixed sources, the AA profile simply won't be a major consideration.

Dieting and AA requirements

As discussed back in Chapter 4, there is evidence for an increased protein requirement when calories are decreased; in that chapter I discussed how daily protein intake should be adjusted during caloric restriction in terms of the total quantity consumed. Continuing with the theme in this chapter, I want to address the issue of whether a specific protein or AA profile might be superior to another while dieting.

The primary mechanism explaining the additional requirement for protein is that the body will be using protein to produce energy (recall from Chapter 3 that skeletal muscle will break down BCAAs to produce alanine which is then used for glucose production in the liver). The more protein being used to provide energy to the body, the less that is available for tissue maintenance (or growth/recovery following training). Hence more needs to be consumed to ensure that requirements for other tissues are met.

Protein can be converted to glucose in the liver via a process called gluconeogenesis and, under conditions of inadequate caloric and protein intake, the body will break down muscle tissue to make glucose. Alanine and glutamine are two of the key players here (33). As with endurance training, they are produced by muscle via the breakdown of the BCAAs. This suggests that proteins high in glutamine or the BCAAs might be particularly useful in sparing muscle loss on a diet. Table 3 on the next page shows the relative proportion of BCAAs and leucine in selected protein sources (33). As mentioned above, few proteins are terribly high in glutamine and supplementation would likely be required to significantly increase glutamine intake.

Recent research has found that high leucine intakes help to maintain blood glucose on a diet when dietary carbohydrates are lowered. That same research found a lower LBM loss with high leucine intakes on a diet. This makes some logical sense, if the body is breaking down skeletal muscle leucine to produce glucose in the liver, providing more dietary leucine should eliminate this need; hence LBM is spared. As Table 3 above shows, whey and casein contain the greatest proportions of both leucine and BCAAs (33).

Table 3: Leucine and BCAA content of selected foods

Food	Leucine (%)	BCAAs (%)
Whey protein isolate	14	26
Milk protein	10	21
Meat protein	8	18
Soy protein isolate	8	18
Wheat protein	7	15

Source for data: Layman DK, Baum JI. Dietary protein impact on glycemic control during weight loss. J Nutr. 2004 Apr;134(4):968S-73S

Related to this, one study that gave wrestlers massive amounts of BCAAs (0.9 g/kg or 63 g/day for a 70kg individual) found a slightly greater sparing of LBM and a greater loss of visceral fat (34). This study is discussed in more detail in Chapter 12 but it's important to note that the protein intake of the wrestlers was not very high, approximately 80 grams per day or less than one-half of what this book would recommend. Whether additional BCAAs would be beneficial given sufficient protein intake in the first place is unknown but seems unlikely.

The high concentration of BCAAs and especially leucine in whey protein makes it attractive on a diet (additionally, the cysteine content may have a beneficial effect while dieting, as discussed in Chapter 12). However, recall from Chapter 2 that casein exerts an anti-catabolic effect due to its slow rate of digestion; this should be beneficial on a diet. Milk protein isolate, containing both whey and casein might be an ideal protein for dieting; empirically it also seems to keep dieters full better than other protein powders.

In one of the only relevant studies I'm aware of, overweight police officers received either whey or a casein hydrolysate along with a reduced calorie diet and resistance training. The casein group lost significantly more body fat (7 vs. 4 kg) and gained more muscle (4 versus 2 kg); this suggests that the slow digestion rate of casein may make it superior on a diet due to its anti-catabolic effects.

With respect to glutamine supplementation while dieting, I'm only aware of one study. It provided a high dose of glutamine (0.35 g/kg or 24 grams for a 70 kg person) to wrestlers on a large caloric deficit, no effect on LBM sparing was found (35). Glutamine would appear to be ineffective as an anti-catabolic compound; ensuring adequate protein and BCAA intake on a diet would seem to be the superior strategy to limit lean body mass loss.

Summary

Arguments are commonly made regarding the superiority of one protein over another in terms of supporting mass gains or sports performance. To examine this topic, I first looked at the issue of maintenance requirements in terms of AAs (rather than just total protein).

With regards to maintaining current levels of body protein, it doesn't appear that any one protein is superior to any other; all high quality dietary proteins are well in excess of even the highest estimated AA requirements for tissue maintenance.

Due to the energetics of endurance training, an increased need for BCAAs, glutamine or both might be postulated. Given the high level of BCAAs in most dietary proteins, as long as endurance athletes obtain sufficient high quality protein (1.7-2.0 g/kg for men and 1.3-1.6 g/kg for women), BCAA requirements should be covered. There is at least one study suggesting that additional BCAA intake might be of use even within the context of sufficient protein intakes.

In contrast to endurance athletes, strength/power training doesn't use protein for fuel. However, glycogen depletion from training could conceivably increase BCAA breakdown (and hence requirements) and increases in lactic acid might conceivably increase glutamine requirements. Outside of those two AAs, the increased protein requirements are generally required to cover the synthesis of new proteins as well as the breakdown of damaged tissue that occurs with training. There is little to no indication that any specific AA profile will be required to support this process. Ensuring sufficient high quality protein (along with sufficient energy intake) would appear to be sufficient.

Limited data examining different types of protein supplementation on strength and size gains tends to support this idea: quantity is more important than protein quality, especially with large intakes of varied protein. At the 2.5-3.0 g/kg (1.1-1.4 g/lb) of protein recommended in this book, with that protein coming from a variety of high-quality sources, within the context of maintenance or higher caloric intakes the AA profile of individual proteins should be completely irrelevant.

During dieting, there is an increased use of protein for energy and this serves to increase overall protein requirements above baseline. Recommendations are given in Chapter 4 and discussed again in Chapter 13. Ensuring sufficient BCAA intake during these periods may exert an anti-catabolic effect with dairy proteins (casein and whey) being high in BCAAs and especially leucine. Milk protein isolate, containing both casein and whey may be an ideal protein during dieting.

7

Meal Frequency

In this chapter, I want to look at the issue of meal frequency and protein intake during the day. Bodybuilders are probably the most obsessive of all athletes when it comes to eating protein (and calories) in small amounts throughout the day (often eating every 2.5-3 hours or so) with the belief that this is necessary for optimal protein utilization, growth or fat loss.

A typically given reason is that the body can only utilize some fixed amount of protein per meal. As discussed in Chapter 2, there may be an element of truth to this idea. An additional goal is to maintain a relatively constant level of amino acids in the bloodstream by ingesting small amounts at frequent intervals.

To one degree or another, most sports nutrition books echo the recommendation for a high meal frequency and there is certainly a significant amount of data to suggest that spreading one's daily calories across more meals has health benefits. However, it's less clear whether a higher meal frequency actually results in differences in how well or poorly the body utilizes nutrients and protein.

First I want to look at some of the potential benefits of a higher meal frequency throughout the day before looking at what little direct data on meal frequency and nitrogen balance exists. Finally I'll look at the topic from a little bit more of a theoretical/physiological standpoint to try and reach some conclusions about optimal meal frequency and protein intake.

Meal frequency and health

There are a number of potential health benefits to spreading out calories throughout the day (although some intriguing recent research, which is outside the scope of this book, is finding that intermittent periods of fasting can also have benefits).

Spreading daily calories out into several meals (usually called a "nibbling" pattern) as opposed to a few larger meals (often called a "gorging" pattern) has been found to improve insulin sensitivity and carbohydrate tolerance (1) and leads to improvements in blood lipid levels (2).

It's important to note that many of the meal frequency studies use a somewhat unrealistic number of meals relative to real-world eating patterns. A typical study might compare a few meals per day to upwards of nine to 17 meals/day. Whether a difference between three to four meals and five to six meals will exist is debatable; some studies find a benefit of six versus three meals, others do not.

There is certainly generally no harm to spreading daily calories into smaller, more frequent meals (outside of possible convenience issues related to meal preparation) but there may not automatically be a benefit from a health perspective.

Meal frequency and body weight/metabolic rate

It is often claimed that higher meal frequencies cause weight loss and that lower meal frequencies (or skipping meals) lead to weight gain. Interestingly, early studies on meal frequency often found that eating more frequently led to weight gain (as opposed to the weight loss benefits often claimed) but this is likely due to the fact that individuals were adding snacks to their normal eating patterns. This is different than taking the same daily caloric intake and splitting it into multiple smaller meals.

Related to that issue, and in contrast to the commonly held belief, meal frequency has no real impact on energy expenditure given an identical caloric intake; eating the same number of calories in a few small meals will cause the body to expend the same number of calories during digestion and metabolism as if they are eaten in many smaller meals (3). The majority of studies have not found a major effect on weight loss (fat loss is less well studied) with varying meal frequency either, at least not when caloric intake is identical.

Rather, it appears that any impact of meal frequency in terms of body weight loss has to do with changes in food intake, probably subsequent to changes in hunger and appetite (3). For example, studies have found that eating more frequently tends to reduce appetite (and thus spontaneous food intake) and this occurs in both lean (4) and obese individuals (5). If eating more frequently keep someone full and they end up eating less total calories, they will tend to lose weight/fat. But this is occurs because they are eating less calories, not because of the meal frequency per se.

In contrast, as noted above, some earlier studies found that eating more frequently caused weight gain, and this was probably because it led to people eating more total food each day. If someone simply adds more food to their normal intake, and this causes them to eat more total calories, this will result in weight/fat gain. But again, this occurs not because of the meal frequency per se but in how the change in meal frequency is affecting total food intake.

Again, this is different than the suggestion to take the same daily caloric intake and spread it across more, smaller meals per day which, generally speaking, has been found to be beneficial in terms of fullness, blood sugar stability, and health.

Practical aspects of meal frequency

Regardless of potential health benefits related to meal frequency, there may be practical reasons to eat more or less frequently. For athletes with high caloric requirements, eating multiple smaller meals may make consuming sufficient calories (and especially carbohydrates) easier to accomplish compared to eating a few larger meals (6). For example, endurance athletes generally have fairly high caloric requirements and cyclists and runners often report eating 8-10 times/day to meet them (6).

Large strength/power athletes can have similarly large food requirements and may need to eat more frequently to avoid the stomach upset that can occur with very large meals. Eating six 1000-calorie meals may simply be easier than trying to force down three massive 2000-calorie meals.

Conversely, individuals with lower caloric intakes (e.g. lighter female athletes trying to reduce body weight) may find that high meal frequencies make individual meals too small to be satisfying or filling. A light female who may be consuming 1200-1500 calories/day would probably find six meals of only 200-250 calories each to be fairly unsatisfying. Utilizing a lower meal frequency, perhaps four meals of 300-375 calories apiece, may be better in this case.

Another issue has to do with the amount of training being done. An athlete who is training 3-4 hours/day (elite cyclists may be on the bike for 4-6 hours/day during which time they do consume nutrients) and who sleeps an additional 8-9 hours/day has a limited amount of time to consume a large number of calories. It can become practically difficult to eat a number of small meals when the total time available for eating is limited; eating fewer larger meals may be the only realistic way to consume sufficient calories to support a heavy training schedule. In contrast, a bodybuilder who may only train 1-1.5 hours/day may find that their training doesn't impact on their meal frequency at all.

A final practical consideration has to do with athletes or individuals who work in addition to their training. Work schedules may or may not allow athletes to fit in 6 small meals per day. So even if such an eating schedule is ideal, it may not be practically possible.

A quick reality check

Athletes often obsess about meal frequency, fearing that missing even a single meal will result in muscle or performance loss. Readers might recall from Chapter 2 and 3 that there are mechanisms such as storage of protein in the gut in place to provide the body with AAs during periods where food isn't available. Additionally, in the very short-term, any proteins broken down when food isn't available tend to come from labile proteins such as liver proteins prior to the breakdown of skeletal muscle.

Perhaps the most direct evidence against the idea that going more than two to three hours without eating causes muscle loss has to do with the digestion rate of most proteins. As discussed in Chapter 2, most proteins digest relatively slowly. Consider, for example, that roughly 40 grams of casein protein may still be releasing amino acids into the bloodstream seven to eight hours later.

The fastest digesting protein is whey, which appears to digest at roughly 10 grams/hour and that only holds for whey consumed by itself under fasted conditions; adding other nutrients slows digestion. In any case, if an athlete consumed 40 grams of whey protein, it would take 4 full hours to digest; almost any other protein would take longer than that. The idea that any time period longer than two to three hours will cause the body to enter a catabolic state simply makes no sense given these facts.

As a singular example, one study found that a relatively modest whole food meal containing 75 grams of carbs, 17 grams of fat and 27 grams of protein was still releasing amino acids and other nutrients (glucose, fatty acids) into the bloodstream at the five hour mark (7); clearly providing more nutrients sooner than that was not necessary to keep the body "fed". Others have suggested that a given meal will continue to exert an anabolic effect on the body for at least five to six hours (8).

An additional point that readers might want to consider is this: over the normal sleeping period of 8 or more hours, the body is generally unfed. Yet an athlete's muscles don't fall off during that time period due to the lack of food intake (I'd mention again that the body stores protein following meals that are released during the fasted period). The idea that a single missed meal, or going longer than three hours without eating will result in gross catabolism of muscle tissue makes no physiological sense.

However, none of the above tells us whether a higher frequency of protein intake would be more or less optimal in terms of muscle mass gain, adaptation to training, or the maintenance of muscle while dieting.

To further address this topic, I first want to examine what little direct research exists on the topic of meal frequency and nitrogen balance.

Meal frequency and nitrogen balance: direct research

While there is no doubt that health and practical issues are important to the issue of optimal meal frequency, none of that information addresses the specific issue of how meal frequency or timing might impact directly on protein utilization, muscle growth, sparing lean body mass loss on a diet, or recovery per se. Surprisingly very little research has been done, most of it is about 30 years old, almost all of it used nitrogen balance and almost none of it has been done in athletes or individuals in training. This makes extrapolations to athletes difficult at best.

For the most part, a majority of studies have found no significant difference in terms of nitrogen balance using a variety of meal frequencies (in these studies, protein and caloric intake was kept constant). However, I'd bring up again that the nitrogen balance method is problematic at best and inaccurate at worst, so it's difficult to draw strong conclusions from the following research.

In one early study, subjects were put on either three or six meals with an identical protein and caloric intake; no difference in nitrogen balance was found (9). The same study also examined the issue of protein distribution throughout the day, studying a group that was given 25% of its protein at breakfast and at lunch and 50% at dinner; again, no difference in nitrogen balance was seen (5). I'll discuss the idea of distributing protein differently throughout the day later in this chapter.

Another study examined the effects of changing meal frequency in women and examined meal frequencies of three, two or nine meals per day; no differences in nitrogen balance were seen (10). In another study, young men were given 1800 calories and 118 grams of protein and either one, three or six meals/day; no differences were seen in terms of nitrogen balance, weight loss or fat loss (11). The same group looked at nutrient utilization under the same conditions and found that one meal was inferior to three or six meals but no difference was seen between three and six meals (12).

In contrast, a more recent study found that six meals/day led to slightly greater weight loss and slightly better nitrogen retention in women on a 1200 cal/day diet compared to only three meals per day (13). Another study, which gave 800 cal/day and 13-15% protein (25-30 g total protein per day) found that five meals was superior to three meals/day in terms of nitrogen balance, but there were no differences in weight or fat loss (14). Please note that 25-30 grams of protein per day is not only below the RDI for protein but also far less than what is required to limit LBM losses while dieting as discussed in Chapter 4.

In possibly the most commonly cited paper on meal frequency and weight and muscle loss, boxers were put on 1200 calories/day which was eaten in either two or six meals per day; while both groups lost muscle the higher meal frequency group lost less (15); this is often cited as proof that a higher meal frequency is superior to lower frequencies while dieting but I find this conclusion to be somewhat flawed.

The main problems are that the study used liquid meals and gave inadequate protein to both groups. Had solid meals (which take longer to digest) or adequate protein been given, I suspect that the results would have been different. Additionally, I consider a two versus six meal comparison somewhat artificial; I suspect that three versus six meals (with solid foods and sufficient protein) would have shown no difference in results.

Beyond the study in boxers above, I'm unaware of any research in training humans that has examined the impact of meal frequency on the adaptations to training. It's possible that research on sedentary individuals simply doesn't apply. Perhaps by increasing the rate of protein turnover, the body needs protein more frequently when individuals are training than when they aren't. These are simply unanswered questions.

However, one very recent study has examined this topic although the data has only been presented in abstract form (i.e. the full paper is not yet available). Thus any results should be considered extremely preliminary. The study could have had methodological flaws that negate the results; without being able to read the full study, there's simply no way to know. However, since it's some of the only direct research into this topic, I want to at least present the results in their preliminary form.

In that study, 33 men and 15 women with at least one year of previous resistance training were placed on either a three or six meal per day diet (16). The three and six meal per day groups were instructed to eat the same number of calories, providing a roughly 300 calorie per day surplus with protein intake set at 1.7 g/kg. Compliance to the diet was measured by self-reporting of food intake.

Both groups performed the same training program over 12 weeks and changes in body composition and body weight were measured. Contrary to what is most commonly believed, the three-meal group gained both more weight and more total LBM.

The reasons for this result are obscure although I'll examine some of the possibilities below. Given the greater total weight (and body fat gain), it may simply be that the three meals per day group simply ate more total calories (the researchers tried to account for this in their statistical analysis). Given the impact on meal frequency and appetite described above, this seems reasonable. As well, although calories were set at identical levels between the groups, self-reported food intakes are often inaccurate; even slight differences in food intake over the length of the study could have significantly impacted on the results seen.

However, as I'll discuss below, there is some evidence that it is possible to eat too frequently which might limit protein synthesis. It's conceivably that the 6 meals per day group crossed this threshold, leading to the observed results. Once again, without being able to read the full manuscript, there's simply no way to know. As well, the results will need to be replicated before any strong conclusions can be drawn.

Summary: Direct research

At calorie balance and given adequate protein, it appears that a higher meal frequency has no real effect on nitrogen balance, at least not in non-training individuals. The exact impact of training on this issue is essentially unstudied; I'll discuss the topic of nutrient timing around training in detail in the next chapter. Which isn't to say that there might not be other reasons of practicality or convenience to adopt one meal frequency over another.

From a practical standpoint, as mentioned above, an athlete with a 6000 calorie per day caloric requirement may find it easier to eat six 1000 calorie meals instead of three massive 2000 calorie meals. Dividing up protein intake across each of the individual meals seems the most logical approach. Athletes with much smaller caloric requirements often find that fewer (but relatively larger) meals are more satisfying and filling.

During dieting, or when protein intake is inadequate, some studies suggest that a higher meal frequency may have benefits in terms of preventing lean body mass loss. Given the apparent impact of different AA levels on protein synthesis and breakdown (discussed in Chapter 2), this makes some sense. Maintaining relatively even levels of amino acids in the blood should help to limit muscle catabolism by decreasing protein breakdown.

Combining a higher meal-frequency with slowly digesting proteins would most effectively accomplish this.

In the boxer study, for example, with only two meals (and liquid meals at that) and insufficient protein to begin with, the body had no source of protein for most of the day; we might expect it to break down body protein under those conditions. With the addition of a third meal, or more total protein, or a slow digesting protein (such as casein or milk protein isolate), the results might have been far different.

How meal frequency affects lean body mass gains with a caloric surplus is essentially unstudied although one preliminary report found that a group consuming three meals per day gained more lean body mass than a group consuming the same calories and protein in six meals per day. As the full paper has not yet been published, the exact reasons for this are unclear.

However, as I discuss below, it's theoretically possible that eating too frequently might actually be detrimental from the standpoint of stimulating protein synthesis.

Optimal meal frequency: A theoretical approach

In Chapter 3, I discussed how eating impacted on both protein synthesis and breakdown following a meal. To briefly recap, an increase in blood AAs primarily stimulates protein synthesis with a much lesser impact on protein breakdown; in contrast, increasing insulin levels appears to primarily decrease protein breakdown with only a small impact on protein synthesis. With that information as background, I now want to examine the topic of meal frequency from a slightly more theoretical standpoint by examining two separate questions:

1. Is it possible to eat too frequently?
2. How long will a typical meal maintain the body in an anabolic state?

By determining a potential maximum and minimum amount of time that should pass between meals, an optimal meal frequency can be developed. As well, I want to examine the idea that different meal frequencies might be optimal under different conditions (i.e. maintenance versus mass gains versus dieting).

Is it possible to eat too frequently?

It's not uncommon to read about bodybuilders or other athletes taking the eat-more-frequently dictum to extreme levels, eating every one to two hours. The idea behind this is the idea that optimal results should occur by maintaining a near continuous influx of nutrients into the body. I imagine if they could find a way to do it, some enterprising athletes would set up a continuous intravenous drip with carbohydrates, amino acids and essential fatty acids.

This may not be a good idea in the first place. Some research, primarily using amino acid infusion, suggests that skeletal muscle can become insensitive to further stimulation of protein synthesis. In one study, amino acids were infused for several hours to 70% over normal levels (17). Protein synthesis increased after roughly 30 minutes and was maintained for the next two hours at which point protein synthesis decreased back to baseline.

Importantly, this decrease occurred despite the maintenance of high levels of blood amino acids. Additionally, there was an increase in urea production (a waste product of protein metabolism), indicating that the excess AAs were simply being catabolized in the liver to be excreted in the urine; that is, those AAs were wasted and never utilized by the muscle.

The researchers took this as a suggestion that there might be a maximum amount of protein synthesis that can occur at any one given time before a "muscle full" situation is reached (18). Perhaps more interestingly, based on the amounts of AAs infused, the researchers estimated that only 3.5 grams of AAs would be required to result in this "muscle full" situation (18). I want to make it very clear that this doesn't mean that 3.5 grams of orally ingested AAs would cause the same effect. Rather, this represented the delivery of 3.5 grams of AAs to the muscle itself.

However, the total amount of dietary protein to achieve this amount wouldn't be huge. Most dietary proteins are roughly 40-50% EAAs, and due to processing in the liver, slightly less than half of the ingested AAs actually make it into the bloodstream. To provide 3.5 g EAAs to skeletal muscle would require roughly 15-20 grams of whole protein over a two hour time span.

Interestingly, other more direct research supports this value. In a study I described in an earlier chapter, subjects received doses of EAAs ranging from zero to 20 g EAAs and protein synthesis was studied (19). In young subjects, muscle protein synthesis was maximized with an intake of 10 g EAAs and there was no further increase with 20 g EAAs. This represents roughly 20-25 grams of whole protein.

Consumed every three waking hours (roughly six meals per day), this would allow for a maximum protein intake of 120 grams per day before skeletal muscle protein synthesis is maxed out. For a 100kg (220 pound) athlete, this is only 1.2 g/kg, lower than even the most conservative estimates discussed in Chapter 4. As discussed previously, this research is a difficult to reconcile with other, much higher recommendations or empirical results.

However, recall from Chapter 4 that dietary protein has more functions for athletes than simply the stimulation of protein synthesis. Although the amount described above might very well maximize skeletal muscle protein synthesis, optimizing the function of other important pathways of AA metabolism would very likely raise requirements even further (20). As well, while excess amino acids may simple be oxidized off, there is evidence that increased AA oxidation is involved in the overall "anabolic drive" of the body.

Finishing up this discussion, in their most recent study, the same group examined the effect on protein synthesis of a variety of doses of infused AAs (21). Infusing AAs at four different ranges, the group saw a similar pattern to their earlier work, an initial increase in protein synthesis followed by a return to baseline despite maintenance of high AA levels.

Additionally, while the lower infusion rates caused a significant increase in protein synthesis, further increases at the higher concentration levels showed smaller additional benefits. Essentially, providing low to moderate amounts of AAs gave the greatest result.

Finally, and perhaps most interestingly, the paper demonstrated conclusively that it was extracellular AA concentrations (rather than the concentration of AAs inside the muscle cell) that were involved in stimulating protein synthesis. The researchers suggested the existence of some type of amino acid "sensor" in the muscle cell membrane that sensed AA levels. The study also suggested that it was the *changes* in extracellular AA concentration, rather than the absolute amounts that were driving the changes in protein synthesis. That is, it was the *change* from lower to higher that had the effect more than the absolute amount of AAs present.

Along with the indication of a "resistance" to further stimulation of protein synthesis, it appears that raising AA concentrations (after a meal) followed by a decrease in concentrations yield the best results. Basically, spacing meals apart and allowing blood AA levels to drop, rather than maintaining AA concentrations at continuously stable levels, appears to have the greatest impact on protein synthesis. Unfortunately, this still gives no indication of how far apart those meals need to be spaced to allow a "resensitization" of the muscle to a subsequent increase in AA concentrations.

Additionally, since it was based on an amino acid infusion, it's unclear how this would relate exactly to the consumption of meals. Between digestion and the hormonal response that occurs with eating, it may very well be that eating protein would yield a different result than what the above research found using AA infusion.

In this vein, it's interesting to look back at the original casein versus whey research that I discussed in Chapter 2. In that study, whey protein showed an initial spike in protein synthesis followed by an increase in amino acid oxidation in the liver, a pattern not dissimilar to the work examined above (22). It seems plausible that once whey had maximally stimulated protein synthesis, the remaining AAs were simply metabolized in the liver.

In contrast, when very small amounts of whey (a few grams at a time) were sipped over a six hour span to mimic the effects of casein, there was no increase in amino acid oxidation (23); however the impact on protein synthesis was also smaller. It may very well be that flooding the body with large amounts of AAs simply overloads the muscle's ability to utilize amino acids, causing the excess to be burned off. This would also be consistent with the fact that the slower protein, casein, actually generated a higher overall gain in leucine in the body compared to whey; by never overloading the body's protein synthetic machinery, overall better results were obtained.

Related to the above research, another group compared the body's use of leucine with subjects either given small hourly meals or three separate meals (24). They found that protein oxidation was decreased (by 16%) in the group given three meals. Essentially, providing amino acids too frequently appears to decrease the body's utilization of those aminos. Rather, having discrete meals where blood amino acid levels first increase (stimulating protein synthesis without overloading the body's ability to utilize AA's) and then decrease for some time (so that muscle can become "sensitive" to the effect of aminos again) would seem to be ideal.

At this point it would appear that eating too frequently (less than every three hours) has no real benefit, and could possibly be detrimental due to the muscle becoming insensitive to the impact of amino acids. It's interesting to note the preliminary report above which found increased LBM gains with three versus six meals per day. Perhaps by spacing the meals further apart, greater stimulation of protein synthesis occurred when protein was eaten.

For the remainder of this chapter, I'll take three hours to represent the minimum amount of time that should pass between meals. Eating more frequently is unlikely to be beneficial and may very well have a negative effect.

How long does a meal maintain the body in an anabolic state?

Having looked at the possibility that eating too frequently might actually be detrimental (or at least not particularly beneficial) given how long a typical meal takes to digest, I want to look at how long a given meal might possibly maintain an anabolic state.

Mentioned above, considering the relatively slow rate of protein and other nutrient digestion, it appears that even a moderate sized meal maintains an anabolic state for at least five to six hours (8). Individual whole food meals are still releasing nutrients into the bloodstream at the 5-hour mark (7). Very slowly digesting proteins such as casein may still be releasing AAs into the bloodstream seven to eight hours after ingestion (22). Considering this research, we might set a conservative limit of five hours as the absolute longest time that should pass between eating some source of dietary protein during waking hours.

Summary: Theoretical examination of meal frequency

It appears that eating too frequently could potentially be detrimental to the goal of gaining muscle mass in that muscle tissue becomes insensitive to further stimulation by amino acids, increasing protein oxidation in the liver. Eating more frequently than every three hours would seem to not only be unnecessary (based on the rate of digestion of whole proteins) but could possibly be detrimental.

Given a moderately sized whole food meal, the body will generally remain in an anabolic state for at least five to six hours (and possibly longer depending on the foods chosen). Conservatively, we might use five hours as the upper limit cutoff for time between meals.

This yields a duration between meals of anywhere from three to five hours. This should keep the body in an overall anabolic state without causing problems related to too frequent or too infrequent consumption of meals.

Full time athletes with time to eat very frequently are probably best served with the higher meal frequency simply to ensure adequate caloric intake. Again, smaller individuals with lower total energy intakes may want to use slightly larger meals eaten slightly less frequently for practical reasons. Similarly, individuals who work jobs and are unable to fit in a meal every three hours needn't worry obsessively about becoming catabolic. A solid food meal containing a high quality protein, carbohydrates, fat and some fiber eaten every five hours will maintain an anabolic state readily.

Protein distribution throughout the day

Related to the topic of meal frequency is the question of whether the day's protein should be spread evenly throughout the day, or if some other pattern of intake might be superior.

As discussed above, one early study examined whether providing 25% of protein at breakfast and lunch and 50% at dinner had any impact on nitrogen balance compared to spreading the protein evenly across the day's three meals; no difference was found (9).

More recent work has examined a dietary strategy called "protein pulse" feeding. With that approach, 80% of the day's protein was given at lunch with only 10% at the other two meals; this was compared to a "spread" pattern where the day's protein intake was distributed evenly across four meals. In elderly women, the "pulse" pattern led to a greater protein gain compared to the "spread" pattern (25). However, in younger women, the "spread" pattern was superior and led to a greater nitrogen balance (26).

There is a substantial and increasing amount of data that putting some amount of the day's protein around training is beneficial, a topic that is discussed in detail in the next chapter. Outside of ensuring adequate protein before, during and after training, there is no real indication that distributing the day's protein in any pattern other than a basic spread pattern is beneficial (again, except possibly for older individuals).

So, for example, take an athlete who will be consuming 200 grams of protein per day with 40 grams of that placed around training. That leaves 160 grams of protein to be evenly distributed across the day's other meals. With a four meal per day frequency, that yields 40 grams of protein per meal; at six meals per day, the athlete would consume roughly 27 grams of protein at each meal.

Is there an optimal intake pattern for different goals?

In the chapter on protein requirements, I mentioned Tipton and Wolfe's contention that any discussion of protein requirements has to be context dependent: that is, the goals of the athlete determine what is optimal in terms of protein intake. While they were talking primarily about total daily protein intake, this idea can be extended to other aspects of nutrition including protein intake throughout the day and how it might interact with specific training goals.

Logically, gaining muscle mass versus maintaining muscle mass at maintenance calories versus trying to maintain muscle mass under conditions of caloric restriction (dieting) are different situations, potentially requiring different optimal intakes of protein, AAs, meal frequency or protein intake pattern. The possibility exists that different patterns of protein intake (in terms of both timing and type of protein) might exist for different goals (27).

For practical purposes, I'm going to consider the following discussion in terms of two different goals: muscle mass maintenance (either at maintenance calories or while dieting) and muscle mass gain. I want to note that most of this discussion will be somewhat hypothetical since little direct research exists to date.

The background for this discussion can be derived from a topic I've discussed previously in the book in terms of how different patterns of protein digestion (i.e. fast versus slow) can influence whole body metabolism differently.

Recapping briefly, large spikes in amino acid concentration appear to stimulate protein synthesis (recall also the infusion data I discussed above) with little to no impact on protein breakdown. In contrast, maintaining constant low levels of AAs appears to reduce protein breakdown with less of an impact on protein synthesis.

Consuming very large amounts of protein at once (as in the protein "pulse" studies discussed above) has an effect similar to a fast protein such as whey, spiking blood amino acids and promoting protein synthesis as well as oxidation (28).

In contrast, spreading protein out in smaller amounts throughout the day has an effect closer to that of casein, inhibiting protein breakdown with a smaller impact on protein synthesis (28).

I'd mention again that, in the original whey versus casein study, reducing protein breakdown via casein had a larger impact on net leucine balance compared to whey. Recall also that adding whey to other food, which had the effect of slowing down digestion, had a similar effect.

Given that data, it may very well be that simply maintaining relatively constant low levels of amino acids (with a spike around training, discussed next chapter) is optimal for all goals. This would be conceptually similar to the strategy of keeping insulin low but stable during the day with a spike around training. This is essentially the strategy that bodybuilders have empirically settled on under all situations: they eat small amounts of protein, carbohydrates and fat throughout the day with a relatievely larger intake of nutrients around training.

With regards to muscle mass maintenance and dieting, there is little to discuss: based on the direct research available as well as the general difficulty in stimulating protein synthesis when calories are reduced, a slow/spread pattern of protein intake is clearly optimal. Maintaining continuous low levels of amino acids throughout the day (in addition to increasing total protein intake) to limit the body's need to mobilize stored body protein from muscle and other tissues should be the goal. A combination of slow proteins combined with evenly spaced meals to keep blood AA levels stable throughout the day would seem to be optimal.

But is this also the optimal pattern for gaining muscle mass? On the one hand there is the suggestive study above where a group receiving three meals per day gained more LBM than a group receiving six per day; as well there is the research suggesting that maintaining constant levels of AAs might cause skeletal muscle to become "insensitive" to further stimulation; increasing extracellular levels of AAs and then allowing them to fall again appears to be superior. Both of these data points suggest that keeping blood AA levels stable throughout the day might not be optimal from the standpoint of muscle mass gains.

Another recent study throws a wrench in the typically held bodybuilder idea that simply maintaining continuous levels of amino acids with frequent meal feeding is optimal (29). In that study, two groups were compared. The first received three whole food meals while the second received the same three meals with an essential amino acid (EAA) supplement in-between. I should note that the study suffered from one huge design flaw: the groups got different amounts of total protein. It should have also tested a group that got 6 whole food meals and the same amount of protein as the EAA supplemented group.

Recognizing that limitation, the study made at least three major observations. The first was that the EAA supplement generated a greater protein synthetic response than the whole meals. The second was that the EAA supplement generated an anabolic response even when given in-between meals. That is to say, the previously consumed meal, which was still digesting when the supplement was given, didn't blunt the effect of the EAA supplement. Finally, the EAA supplement didn't blunt the anabolic response to the meal. Of course, the study didn't examine what impact this would actually have in the long-term on muscle mass gains but is interesting nonetheless.

This study suggests that a potential pattern at least worth experimenting with for athletes seeking maximal muscle mass gains would be to alternate between slower digesting meals with faster acting sources (perhaps a whey protein drink or an EAA supplement) throughout the day (25).

It also plausible that a combination of slow and fast protein sources at a given meal could give the best of both worlds: a spike in AAs to stimulate protein synthesis followed by a slower increase to inhibit protein breakdown. Preliminary data that I discussed back in Chapter 2 supports that idea as well although it was being primarily applied to protein intake following resistance training. It's interesting to note that old school bodybuilders often consumed copious amounts of milk to gain lean body mass as milk protein is a mixture of whey and casein.

8

Nutrient Timing Around Workouts

Although most early research examining the role of dietary protein for athletes focused solely on the total amounts ingested, an area of keen interest recently has to do with the timing of protein (and other nutrients) around training.

Some researchers have suggested that the timing of protein around resistance workouts may be more important in terms of muscle growth and adaptation to training than the total amount of protein consumed (1,2); additionally, various lines of research (discussed below) suggest a role for protein before, during and after training for endurance athletes.

In any case, proper nutrient timing around training has the potential to impact on several aspects of athletic performance. This includes performance during training or competition, recovery between workouts, and the promotion of optimal adaptations to training.

Regardless of the athlete or sport, nutrient timing around training encompasses 4 specific phases: the pre-workout meal (generally consumed 1-4 hours before a workout), immediate pre-workout nutrition (immediately before training to 30 minutes before training), during workout nutrition (anything consumed during the workout itself), and post-workout nutrition (nutrients consumed immediately after training). I'll discuss each in some detail below and provide recommendations in terms of both quantity and type of nutrients that should be consumed. I also want to address some existing controversies and questions regarding the research on this topic as well.

The interaction between carbohydrate and protein

While this book is meant to be about protein, you'll find that in this chapter I also discuss other nutrients, primarily carbohydrate. In terms of nutrient timing around workouts, it becomes difficult to separate out the impact of protein and carbohydrate, as they tend to have additive and interacting roles. This necessitates them being discussed together.

It's safe to say that the majority of early research into the topic of nutrient timing dealt with endurance athletes, as that was the main group studied by exercise physiologists for many years. Research has examined pre-, during- and post-workout nutrition for this group although most of it has focused solely on carbohydrate intake.

More recently, the role of protein around endurance training has begun to be appreciated and more research on resistance training has begun to appear. Additionally, there is some overlap between different sports and some of the data on endurance athletes can be applied to other groups such as strength/power athletes or bodybuilders.

I bring this issue up as it's altogether too easy for endurance athletes to think that all they need to worry about is carbohydrates around training to the exclusion of protein. In contrast, strength/power athletes often assume that all they need to worry about is protein intake often ignoring carbohydrate needs. The nature of high level sports training makes that type of attitude misguided for a number of reasons.

Both endurance athletes and strength/power athletes synthesize proteins in their muscles in response to training. As I've mentioned previously, the proteins made are different, strength/power training stimulates the synthesis of contractile proteins while endurance training promotes increase in mitochondria and the enzymes involved in energy production. Regardless, promotion of protein synthesis, as part of an overall adaptation to training is important for both groups. As you'll see below, endurance athletes seeking optimal adaptations to training and recovery need to worry about proper protein intake.

With a few exceptions, most strength/power athletes do more in training than just weight room or power training. Various types of metabolic work is often done with many sports such as strongman, wrestling or mixed martial arts using large amounts of glycogen during training; this makes proper carbohydrate intake important. Additionally, many athletes train more than once per day; refilling muscle glycogen between workouts becomes crucial for both training and recovery.

My point being that sufficient carbohydrate and protein intake around training is important for almost all athletes. Which isn't to say that each group should receive identical recommendations for carbohydrate and protein intake around training. The specifics of different types of training affect whether relatively more carbohydrate or protein should be consumed.

In general, and this will be reflected in the guidelines below, strength/power athletes will tend to need relatively more protein and less carbohydrate around training; endurance athletes will generally consume more carbohydrate and less protein around training.

The pre-workout meal

The pre-workout meal encompasses anything eaten by an athlete 1-4 hours before a workout or competition. The primary goal of this meal (beyond being part of an athlete's daily nutrition) is to ensure optimal levels of muscle and liver glycogen as well as blood glucose; this is all meant to optimize performance during the training session or competition. Protein, fat and fiber should generally be part of this meal as it really represents a normal meal in the athlete's daily regimen. In general, the pre-workout meal should not be skipped if optimal performance is the goal. There are a few exceptions to this, however.

A major exception to this is an athlete who is on the edge of not making weight for a competition and who must weigh in with minimal food in their system. Assuming that there is sufficient time between weigh-in and competition, an athlete in this situation should eat something immediately after weighing in to ensure optimal blood glucose levels and hydration for competition.

Additionally, endurance athletes occasionally perform lower-intensity endurance training in a fasted state in an attempt to improve their body's utilization of fat for fuel and drive further endurance adaptations. Similarly, bodybuilders and others often perform low-intensity aerobic activity (generally first thing in the morning) without eating in an attempt to lose fat; whether fasted cardio actually generates greater fat loss is debatable and beyond the scope of this book. However, very low intensity (aerobic) workouts generally don't require as much attention to pre-workout nutrition as higher intensity workouts such as weight training or higher intensity conditioning workouts.

The final exception, and I'll discuss this further below, has to do with athletes who must train first thing in the morning. It's often unrealistic to consume a whole food meal one or more hours before early morning training without having to get up so early as to interrupt sleep and recovery. Depending on how long an athlete has between waking up and their first training session of the day, a true pre-workout meal may or may not be realistic. With an hour or more before training, a small solid or liquid meal may be possible. With less than an hour, only immediate pre-workout nutrition (discussed in the next section) may be viable.

In terms of pre-workout meals, most of the research done to date has focused on endurance athletes and has examined the impact of carbohydrate and/or dietary fat on performance with little to no research examining protein consumption (3). Little has looked at the impact of the pre-workout meal on strength/power athletes.

A mixed meal, containing carbohydrate, protein, fiber and some fat is generally ideal here with the amount of food eaten being larger the further away from training the meal is consumed. An athlete eating 4 hours before a training session or competition should consume more food than one having their pre-workout meal 1-2 hours before a workout. This is simply to avoid stomach upset and provide adequate time for digestion. Additionally, the closer to training that the pre-workout meal falls, the more easily digesting foods that should be consumed.

I want to mention that athletes differ in how well they perform with food in their stomach, some seem to do well with a full stomach while others do not; a lot of this depends on how close to training the meal is eaten and liquids may work better if the meal is eaten very close to the workout or competition. Different pre-training and pre-competition meals should be tested to determine which foods give the best results in terms of performance, energy levels, etc.

As mentioned above, athletes who are forced to train first thing in the morning (who may not have several hours to eat beforehand) may or may not be able to consume a true pre-workout meal at all. With an hour or more prior to training, a pre-workout meal can generally be consumed; athletes will probably want to focus on liquid nutrition to avoid stomach upset. Even a small amount of carbs and protein (i.e. a glass of skim milk or a carbohydrate/protein drink) can boost blood glucose and provide amino acids to improve both performance and adaptation to training. In the case of an early morning workout where there's no time for even a small pre-workout meal, an athlete should ensure appropriate immediate pre-, during- and post-workout nutrition.

Consumption of anywhere from 1.1-4.5 g/kg (0.5-2 g/lb) of carbohydrates 1-4 hours before workouts has been recommended to optimize muscle and liver glycogen levels. As mentioned above, the major determinant of how large the meal can or should be is the length of time between the meal and the workout. So, a meal eaten 3-4 hours before training might contain a full 3.0-4.5 g/kg (1.5-2 g/lb) of carbohydrates, a meal 1 hour before workout might contain 1.1-2.2 g/kg (0.5-1 g/lb) of carbohydrate.

It's important to note that these recommendations were originally determined for endurance athletes who are typically doing very extensive (multiple hours) of training; strength/power athletes or endurance athletes doing shorter workouts are unlikely to require as much carbohydrate in the pre-workout meal. Using the lower ranges of recommendations may be more appropriate in this situation.

An additional issue to consider in deciding on optimal amounts of pre-workout nutrients is the nature of the workout. As the volume or intensity of the workout goes up, the pre-workout meal should become larger; as workout intensity or volume decreases, the pre-workout meal can decrease in size. Again, a possible exception would be long-duration low-intensity aerobic training (i.e. a 1-2 hour easy spin by a cyclist) but even there, blood glucose and liver glycogen can become limiting and ensuring optimal stores with a proper pre-workout meal can help to avoid bonking and performance loss.

Along with carbohydrates, protein should also be consumed although few guidelines exist. In fact, I'm only aware of one study that has looked at the topic at all and it examined the impact of BCAAs more than protein per se. That study compared the impact of BCAAs (approximately 10 grams of BCAAs along with 12 grams of milk protein, 30 grams of carbohydrate and a trivial amount of fat) to the same drink without BCAAs and more milk protein. The drinks were given 90 minutes before a one- hour run. The consumption of the BCAAs prevented testosterone and GH from falling following the run (4).

For the most part, the pre-workout meal should simply be considered a normal meal in the first place. So it should contain similar amounts of protein to any other meal consumed during the day. How much protein that will end up being will, of course, depend on the athlete's total daily protein intake. The greater the daily protein intake, the larger the amount of protein that will be consumed at this meal.

On average, the pre-workout meal might contain anywhere from 20-40 grams of protein although much smaller or larger athletes might find that this meal will contain more or less than this. I'd note that, on average, endurance athletes would typically be consuming less total protein and relatively more carbs at this meal compared to a strength/power athlete.

This primarily reflects the overall difference in daily diet for the different athletes due to the different physiologic demands of their sport. In addition, some amount of fat, depending of course on daily caloric requirements and intake would also be consumed here, along with a fiber source such as fruits or vegetables.

As mentioned above, in addition to being smaller overall, the closer this meal is eaten to the workout, the more easily digested it should be. Fat intake would be lowered and slowly digesting high-fiber items would be omitted to avoid stomach upset. A meal consumed 3-4 hours before training should pose no real problems in this regards but a meal consumed only an hour before training could cause problems due to the presence of food in the stomach during intense training.

Immediate pre-workout nutrition

The second phase of nutrient timing is immediate pre-workout nutrition; this encompasses anything consumed from roughly 30 minutes before to immediately before training or competition. Any nutrients consumed during the warm-up period might also be included in this phase. The purpose of this phase is similar to that of the pre-workout meal: to ensure optimal blood glucose levels. Hydration is also a goal here and, as you'll see below, consuming protein immediately before training may be superior to post-workout protein intake in terms of driving protein synthesis.

As usual, most of the research in this area has been done on endurance athletes and has focused primarily on carbohydrate intake. Recently, one or two key pieces of research on resistance training have appeared.

As mentioned above, one of the purposes of immediate pre-workout nutrition is to ensure optimal blood glucose levels; however, this topic is not without some controversy. Early research suggested that consuming carbohydrate immediately prior to training hurt performance by causing blood sugar to crash due to an insulin spike but this finding was hardly universal: all other studies either found that performance is improved or unchanged with pre-workout carbohydrate intake (3).

Additionally, consuming carbohydrates during the warm-up prior to cycling prevented the normal increase in insulin and drop in blood glucose (5). Athletes who are prone to crashing blood sugar may want to delay the consumption of their immediate pre-workout nutrients until they are already into their warm-up.

Little work has examined the impact of protein prior to endurance training although some work has looked at BCAA supplements prior and/or during endurance training from a performance standpoint. The results from these studies sometimes find a benefit but the results are highly mixed; some research suggests that BCAA supplements could actually harm performance. This research is discussed in some detail in Chapter 12.

I'm unaware of any research which has examined the impact of protein immediately before endurance training although some recent work given a combination of protein and carbohydrates before and during workouts, however. That research is discussed below in the section on during workout nutrition.

With regards to strength/power athletes, some research has suggested a benefit of pre-workout carbohydrates (6) and a recent study found that consuming 1.0 g/kg (0.45 g/lb) carbohydrates before and an additional 0.5 g/kg (0.22 g/lb) of carbs every 10 minutes during workout significantly reduced the decrease in muscle glycogen from training (7). Limiting muscle glycogen depletion can be important from a performance standpoint, especially for athletes who train twice daily.

Looking at protein, recent research suggests that pre-workout protein consumption might impact on protein synthesis after training to a greater degree than the same nutrients consumed afterwards. One study found that a combination of 35 grams of sucrose and 6 grams essential amino acids (EAAs) taken immediately prior to resistance training had a greater impact on post-workout protein synthesis than the same drink taken immediately afterwards (8). I'd note that most of the difference was due to a single subject who showed a massive response to the pre-workout nutrients so the effect may not be universal.

Interestingly, the response was not seen when whey protein was consumed immediately before workout, possibly due to a slower digestion time compared to the EAAs (9). The researchers suggested that perhaps consuming whey protein 30 minutes before training might allow sufficient time for digestion to get the benefits of pre-training supplementation. Consuming a protein hydrolysate immediately prior to training might also have the same impact (10). My feeling is that consuming a fast protein such as whey (or soy) 30 minutes prior to a training bout should allow for sufficient digestion and AA delivery to promote protein synthesis post-training.

Another recent study, which I'll discuss in more detail below, examined the impact of a dextrose, whey protein and creatine drink consumed immediately before and after training on muscle growth; this was compared to a group who received the identical supplement at other times of the day but not around their workout (11). The supplement contained approximately 0.40 g/kg (0.18 g/lb) of whey protein, 0.43 g/kg (0.20 g/lb) of dextrose, and 0.07 g/kg (~0.03 g/lb) of creatine with trace fat before and after training.

For a 100kg (220 pounds) lifter, this provided 40 grams of whey protein, 43 grams of carbohydrates, and 7 grams of creatine both before and after training. That study found that lean body mass gains were significantly higher compared to the group who took the same supplement at other times of the day. Interestingly, there was also a slight fat loss in the group consuming the supplement around training compared to the group consuming it at other times of the day.

It's currently unknown whether protein/AAs taken prior to endurance training would have the same impact on post-training protein synthesis or adaptation but endurance athletes may wish to consider adding a small amount of protein to their immediate pre-workout drink in an attempt to promote greater adaptations and recovery following training. As I'll discuss below in during workout nutrition, this may have the added benefit of limiting protein breakdown during exercise.

With the exception of the two studies mentioned above, one using EAAs and the other using whey protein prior to training, research has not systematically examined different types of protein prior to training. Practically speaking, to avoid stomach upset, a fast protein such as whey protein or soy is the best option. Using a rapidly digesting carbohydrate source such as dextrose or sucrose makes the most sense here as well although, again, susceptible athletes should watch for signs of a blood glucose crash. Due to its slower digestion time and tendency not to raise insulin, fructose (or fruit) may be a viable option here as well.

Prior to strength/power training I recommend an intake of 0.3-0.5 g/kg (0.13-0.22 g/lb) of carbohydrate with an equal amount of protein roughly 30 minutes before training (to allow time for digestion). If an EAA supplement is used roughly one half as much (0.15-0.25 g/kg or 0.06-0.11 g/lb) would be appropriate and it can be consumed closer to training. This should optimize blood glucose levels and provide amino acids during the training session. Three to five grams of creatine could be added to this supplement as well. For a 100 kg athlete, this would equate to 30-50 grams of carbohydrate with 30-50 grams of whole protein. That amount of protein will provide roughly 15-25 grams of EAAs and 8-12 grams of BCAAs. If EAAs are used, 15-25 grams with carbs would be an appropriate intake.

Before endurance training, a carbohydrate intake of 1.0 g/kg (0.45 g/lb) with perhaps 0.15-0.25 g/kg (0.06-0.11 g/lb) of protein should be sufficient. For a 70 kg (154 lb) athlete, this would provide 70 grams of carbs and 10.5-17.5 grams of protein. This will provide 5-8 grams of EAAs and 2-5 grams of BCAAs.

To avoid stomach upset, both fat and fiber should be avoided in the immediate pre-workout meal. Immediate pre-workout nutrient recommendations are summarized below in Table 2.

Table 2: Immediate Pre-workout nutrition recommendations

	Protein	Carbohydrate
Endurance	0.15-0.25 g/kg	1.0 g/kg
Strength/power	0.3-0.5 g/kg*	0.3-0.5 g/kg

* Half of this value, or 0.15-0.25 g/kg of EAAs could be used instead

During workout nutrition

The third phase of workout nutrition deals with any nutrients consumed during the workout period itself. As mentioned above, there is potential overlap with the immediate pre-workout nutrition phase depending on whether or not the warm-up period is considered as being part of the workout.

For years, the primary goal of during workout nutrition, especially for endurance athletes, was aimed at maintaining blood glucose levels along with hydration; this was especially crucial for longer workouts or competitions. More recent research has found that proper during workout nutrition may improve performance in shorter endurance events as well as benefiting strength/power athletes during their workouts. As well, while most of the research concerned itself with carbohydrate intake, more recent research has found that consuming small amounts of protein during workout can limit muscle breakdown, may improve performance and have other benefits for both endurance and strength/power athletes.

As mentioned above, during endurance training, athletes have two primary goals in terms of nutrition: maintaining blood glucose levels and staying hydrated. As far as blood glucose was concerned, early research suggested that carbohydrates were only really needed during training sessions or races lasting longer than 90-120 minutes. More recent research has suggested a performance benefit for events of an hour or less as well (12).

Maintaining adequate hydration is primarily a function of consuming sufficient fluids during training (32-36 oz per hour being near the maximum absorbable amount) although adding carbohydrate in moderate amounts improves uptake in the stomach; adding small amounts of sodium and potassium is mainly done to improve taste. Commercial hydration drinks (such as Gatorade and Powerade in the US) are generally designed along these lines: providing sufficient amounts of carbohydrate and minerals to ensure both adequate fluid and carbohydrate intake.

It has also been shown that consuming carbohydrates during endurance training helps limit the increase in cortisol and impairment of immune system function along with decreasing oxidative stress and the need to utilize protein for fuel (13,14,15). With the exception of training done explicitly under fasted conditions mentioned above or very low intensity activity, I feel that carbs should always be ingested during endurance training.

There has generally been thought to be a limit to how much carbohydrate can be consumed and effectively utilized during endurance training, with the gut showing maximum absorption rates of 30-60 grams/hour of glucose. However, by using a combination of glucose and fructose (which use different intestinal transporters), 20-50% more carbohydrate can be consumed and utilized without stomach upset (12). This leads to greater use of carbohydrate during exercise and may improve performance by sparing muscle glycogen use.

I'd note that while a majority of research has used simple sugars (such as glucose or sucrose) during exercise, longer chains of glucose such as maltodextrin or maltose are also effectively utilized during exercise (16). Additionally, a combination of maltodextrin and fructose has been shown to promote higher carbohydrate utilization rates than maltodextrin alone (17).

A combination of glucose and fructose also improves fluid absorption from the gut which may help with the maintenance of hydration (18). By using a combination of glucose and fructose (note that sucrose is half glucose and fructose), a carbohydrate intake of 45-70 grams per hour could be achieved during endurance training; as mentioned above, this would be mixed in 32-36 oz of fluid.

With regards to resistance training, while it hasn't traditionally been thought that carbohydrates during (or before) training were particularly important, recent research has found that consuming 1.0 g/kg (0.45 g/lb) of carbohydrate limits muscle glycogen depletion, keeps insulin higher and reduces cortisol levels. The same amount of carbohydrate may also improve performance (especially for training sessions lasting an hour or more), and leads to higher levels of blood glucose following training (19). All of these effects could have an overall positive impact on recovery and strength and muscle mass gains (discussed in the next section).

But what about protein during training? One early study provided a 6% carbohydrate solution (sucrose) with 6 grams of EAAs during resistance training (20). It found that insulin stayed higher, cortisol was lower, and muscle breakdown during training was decreased; this might be expected to improve recovery and skeletal muscle gains in the long-term. Like many of the studies on this topic, it's important to realize that it was done in the fasted state and the results may have been different if the subjects had eaten before the workout.

Another recent study provided approximately 62 grams of carbohydrates with 15 grams of protein (the type was not specified) around and during resistance training. One third of the drink was consumed 30 minutes before training with another third given during training and the final third consumed immediately after training. Although cortisol was unaffected, insulin levels were higher and the study found a significant decrease in protein breakdown during training as well as 24 hours later (21).

In terms of endurance training, while early research focused exclusively on carbohydrate intake, recent work has begun to find a number of potential benefits of consuming both carbohydrates and protein during endurance exercise. These include improved performance, decreased muscle damage, improved recovery, and better performance at subsequent workouts (22). As well, adding small amounts of protein to carbohydrate may increase fluid absorption from the stomach, further helping with hydration status during exercise.

A variety of work has been done to date examining differing amounts of both carbohydrate and protein. Typically a ratio of approximately 4:1 carbohydrates to protein (e.g. 40-60 grams of carbohydrate to 10-15 grams of protein) has been used. At least three published studies have found that this can improve performance during endurance exercise (13) while one found no effect (23). Interestingly, this latter study used a larger amount of protein (20 g/hour) and this may have been in excess of optimal amounts (13). It's currently unclear how adding protein to carbohydrate during endurance training improves performance (22,24).

Of additional benefit, several studies have found that adding protein to carbohydrate during training decreases protein breakdown and measures of muscle damage, improving performance at subsequent training sessions (22, 25). A carbohydrate/protein gel containing providing 40 grams of carbohydrates per hour and 10 grams of protein per hour was recently shown to both improve performance during cycling as well as decrease markers of muscle damage (26).

Given the potential contribution of AAs to energy production during long-duration endurance training, the above research makes some sense. Providing dietary AAs would be expected to limit the need to break down muscle protein during training; this should improve overall recovery.

Typically, the studies to date have used whey protein during training although recent work has also used a casein hydrolysate (22). No studies have systematically compared different types of protein during endurance training but, from a practical standpoint, using a quickly digesting protein during training makes the most logical sense to avoid stomach upset.

As a final issue relating to during workout nutrition, I should probably talk somewhat about BCAA supplementation, although the topic is discussed in much greater detail in Chapter 12. The BCAAs, through a variety of pathways, have been found to have potential benefit during training; they may decrease fatigue (although recent research has called this into question) and some coaches feel that high dose BCAA intake during strength/power training may improve adaptation. There is certainly some research to support this idea although none has compared, for example, high dose BCAA intake during training to the consumption of carbs/whey protein as I'm discussing in this chapter.

As well, while early studies suggested a positive effect of BCAAs on endurance training performance, evidence of impaired performance was also found due to the production of excess ammonia (a potential cause of fatigue) during training. In contrast, the consumption of whey proteins during training doesn't increase ammonia in this fashion and may be a better strategy overall.

In any case, it seems unlikely to me that anything but pharmacological doses of BCAAs (20-40 grams are often recommended) taken during training would confer a benefit. As well, athletes following the full recommendations I'm making in this chapter for pre-, during- and post-workout nutrition will already be obtaining a fairly large amount of BCAAs (and EAAs) around training without the need to supplement.

I should also mention that some strength/power athletes, especially those whose workouts involve long rest intervals between sets will mention blood sugar crashing when they consume large amounts of high GI carbohydrates during training. This is especially true if they use dextrose or glucose. Using sucrose or something akin to Gatorade (which is a combination of glucose, sucrose and fructose) during training can often eliminate this problem as the digestion speed is slower. However, for those athletes who still have issues, avoiding carbohydrate intake during training may be the only option. Consuming a dilute solution of whey protein or BCAAs may be preferred.

At this point it looks like an hourly intake of 30-60 grams of rapidly digested carbohydrate such as dextrose or sucrose (or a mix of dextrose/glucose and fructose) with 8-15 grams of an easily digesting protein such as whey can be beneficial for both strength/power and endurance athletes.

Using a combination of dextrose/glucose and fructose would allow endurance athletes to consume up to 70 grams of carbohydrate per hour, increasing carbohydrate oxidation rates and possibly sparing muscle glycogen. Endurance athletes may wish to experiment with progressively larger amounts of carbohydrate to find their individual tolerance for carbohydrate while avoiding stomach upset. I'd mention again that table sugar (sucrose) is one half glucose and fructose.

As has been the case throughout this chapter, due to the differences in demands between activities, strength/power athletes will generally be consuming slightly less carbohydrate and slightly more protein than endurance athletes who will consume relatively more carbohydrate and relatively less protein during training.

Table 3 below shows recommended amounts of nutrients to be consumed during training. The amounts listed below would be mixed in 32-36 oz of water and sipped throughout the workout at a rate of 8-9 oz every 15 minutes.

Table 3: During workout nutrition recommendations

	Protein	Carbohydrate
Endurance	8-15 g/hour whey protein	30-60 g/hour dextrose alone OR 45-70 g/hour dextrose plus fructose
Strength/power	12-15g/hour whey protein	30-45 g/hour dextrose or sucrose

Note: Carbohydrates such as maltodextrin, maltose or sucrose can be used in place of dextrose if desired or better tolerated by the athlete. As well, soy could be used in place of whey.

Most athletes find that the addition of small amounts of potassium and sodium improve flavor and make them want to drink more compared to unflavored drinks. A mixture of a commercial sports/hydration drink such as Gatorade or Powerade along with whey protein mixed in 32-36 oz of water gives a close approximation to the studies done to date. The amount of Gatorade/Powerade powder and whey protein can be varied to achieve the recommended intake levels in Table 3 above.

For example, a strength/power athlete would mix 2-3 tablespoons of Gatorade powder (providing approximately 30-45 grams of carbs) with just over half of a standard scoop of protein powder (providing 12-15 grams of protein). An endurance athlete would put 4 tablespoons of Gatorade with roughly 1/2 scoop of whey to obtain approximately 60 grams of carbs and 10-15 grams of protein.

Post-workout nutrition

The final phase of around workout nutrition is post-workout nutrition; this phase refers to any nutrients consumed immediately after training until perhaps 1-2 hours later. At some point, of course, the post-workout period becomes the normal daily eating plan but this really isn't that important a distinction as far as I'm concerned.

What is important is that what is consumed immediately after training has the potential to impact significantly on various aspects of training including glycogen resynthesis, protein synthesis, recovery, adaptation to training and subsequent workout performance.

I'm going to divide the work in this area up into two distinct categories: glycogen resynthesis and protein synthesis although, as you'll see, there is overlap in the recommendations for both.

Glycogen resynthesis: Introduction

Glycogen is the storage form of carbohydrate found in the muscle and liver. Maintenance of optimal levels of muscle glycogen is important for performance and affects a number of aspects of metabolism relevant to athletes. Glycogen levels can impact on gene expression, protein breakdown, protein synthesis, and affect whole body metabolism (27).

As has typically been the case, most of the early work on glycogen depletion and resynthesis came from studies of endurance athletes although more recent research has examined resistance training as well. Decades of studies determined a significant number of factors related to glycogen resynthesis following training. These issues could be separated into issues of amount, type and timing of carbohydrate.

I want to make it clear that the research I'm going to discuss is only directly applicable to endurance athletes engaged in daily, near-exhaustive training although some sports nutritionists have uncritically tried to apply it to non-endurance athletes.

A cyclist performing 3-6 hours/day on the bike is going to require far more carbohydrates (and total calories) than, say, a mixed martial artist performing 45-60 minutes of high intensity work as part of their conditioning several times per week. Powerlifters and Olympic lifters, who typically work in a low repetition range may not deplete much muscle glycogen at all, while a bodybuilder working in a medium-high repetition range, especially if they are doing a large number of sets, may be getting significant amounts of glycogen depletion. The bodybuilder will have proportionally higher carbohydrate requirements. Strongman competitors who tend to do a great deal of strength endurance training will also be depleting muscle glycogen. Athletes in many team sports often do large amounts of resistance training along with conditioning work and will have proportionally higher carbohydrate and calorie requirements.

My point is that there is no one size fits all recommendation for carbohydrate intake that can apply to all athletes. This applies both to total carbohydrate intake in addition to any nutrients consumed around training. Differences in training volume, intensity, and the primary energy system used all play a role in determining carbohydrate requirements.

In any case, glycogen repletion requirements can be divided into two categories depending on how long the athlete has between training or competition: when the duration between training bouts is 6-8 hours or less (which is common in athletes training twice daily), the requirements for optimal glycogen replenishment are far different than when there are at least 24 hours between training sessions or competition (common among recreational athletes or those training once daily or less frequently).

Glycogen replenishment with at least 24 hours between workouts

When an athlete has at least 24 hours to replenish glycogen, it turns out that the *amount* of carbohydrate consumed is the primary determinant of whether or not normal glycogen levels can be achieved (28); the *type* of carbohydrate or *timing* of intake aren't as important. As long as amounts in the range of 10-12 g/kg (4.5-5.5 g/lb) of carbohydrates are consumed during the 24 hours between workouts, glycogen can be restored within that time span.

Under these conditions, timing doesn't seem to matter much. Whether 2 or 7 meals are fed, or the carbohydrates are given as four large meals or 16 small snacks given an hour apart, the glycogen levels achieved are the same (28). The same holds true for the type of carbohydrate. As long as extremes of diet (i.e. all fructose) are avoided, the type of carbohydrate (high versus low glycemic-index or simple versus complex carbohydrate) is also fairly irrelevant.

While 10-12 g/kg (about 4.5-5.5 g/lb) of carbohydrate may be necessary for endurance athletes, this is probably excessive for most strength and power athletes. Roughly half that amount or 5-6 g/kg (2.2-2.7 g/lb) should generally be sufficient and may be excessive depending on total training volume and intensity. Once again, much of this depends on the specifics of the sport and the type and amount of training being done.

By monitoring performance and changes in body composition, athletes can eventually determine whether their intake of carbohydrates and total calories is insufficient, excessive or just right. But for the most part, only endurance athletes involved in daily exhaustive (2-6 hour) training sessions will need to target the full 10-12 g/kg of carbohydrate on a daily basis.

Glycogen replenishment with 8 hours or less between workouts

In contrast to the situation where an athlete has 24 hours or more to replenish glycogen stores, when the time between sessions is only a few hours (which may be the break an athlete gets when training or competing twice per day) things get more complicated. Issues of type, timing and amount of carbohydrate end up significantly impacting how well or poorly glycogen can be resynthesized.

Summarizing the research on this topic, it was found that glycogen storage could be maximized by consuming a large total amount of carbohydrates in smaller amounts immediately after training and for the 4-6 hours after training. A carbohydrate intake of 1.0-1.85 g/kg/hour (0.45-0.8 g/lb/hour) consumed in smaller drinks every 15-60 minutes will maximize glycogen storage following exercise (29).

For example, a 70 kg athlete would consume from 70-130 grams of carbohydrates per hour with drinks being consumed every 15-60 minutes. Let's assume he's going to consume a middle value of 100 grams of carbohydrate per hour. If he took one drink per hour, he'd consume the full 100 grams of carbs. If he consumed carbs every 30 minutes, that would be 50 grams in each drink. Or he could consume 25 grams of carbohydrate every 15 minutes.

This would be continued during the 4-6 hours following training to maximize glycogen resynthesis between training bouts or competition. Athletes who do not wish to consume this many carbohydrates may be able to achieve high levels of glycogen synthesis by combining smaller amounts of carbohydrates with very insulinogenic proteins such as whey; this topic is discussed below.

It is important that carbohydrates be consumed very soon after the completion of training; a delay of several hours led to 50% less glycogen resynthesis compared to consuming those carbs immediately following training. This is probably where the idea of a "window of opportunity" for recovery came from. Once again, this research was performed on endurance athletes not strength/power athletes. As mentioned above, many strength/power athletes engage in training twice per day and ensuring optimal glycogen replenishment between bouts of training is important for those types of athletes as well.

In terms of the type of carbohydrate, when rapid glycogen synthesis is the goal, higher glycemic index carbohydrates work more effectively than lower glycemic index carbs, primarily due to increased insulin levels. Glucose and sucrose are more effective than pure fructose at refilling muscle glycogen.

Other forms of carbohydrate may be even more effective when very rapid glycogen replenishment is desired. Recent research has found that a high molecular weight carbohydrate replenishes muscle glycogen more quickly in the first 2 hours after exercise compared to glucose (30). The same compound was recently shown to improve performance in a time trial undertaken 2 hours following an exhaustive exercise bout (31).

This product is sold commercially under the trade name Vitargo or in bulk as waxy maize starch. Athletes who need to replenish glycogen at maximum rates may want to experiment with this product following workouts when they only have a short period to resynthesize glycogen.

Failing that, using glucose/dextrose or maltodextrin immediately post-training followed by other high glycemic index carbohydrates is the most effective way to replenish glycogen rapidly on days when two training sessions must be performed with only a short-break between them. A small amount of fructose can also be beneficial in order to refill liver glycogen, but the primary carbohydrates should consist of dextrose/maltodextrin or glucose as those are preferentially used to refill muscle glycogen.

Carbohydrate and protein post-workout: Improved glycogen synthesis

While the majority of early research focused solely on carbohydrate intake and its effect on glycogen resynthesis, this changed in 1992 when a seminal paper found that adding protein to carbohydrate increased glycogen resynthesis following exercise (32). In that study, male cyclists rode for 2 hours and then received either 112 grams of carbohydrate, 40 grams of protein, or a combination of the two.

The carb/protein combination led to significantly larger amounts of glycogen synthesis after four hours compared to either carbohydrate or protein by itself. Adding protein or certain amino acids increases the insulin response to carbohydrate and this is most likely what caused the difference in glycogen storage (33,34).

A number of similar studies followed, some found that protein added to carbohydrate increased glycogen storage while others did not. Without detailing each (see references 28 or 29 for review), the following pattern developed: when carbohydrates were fed in large quantities at frequent intervals, protein provided no additional benefit.

However, when carbohydrate intake was below a certain threshold of intake (approximately 1.2 g/kg/hour or 0.5 g/lb/hour) or the feedings were more than an hour apart, additional protein enhanced muscle glycogen storage. Once again, increased insulin release from the protein appears to be the primary mechanism and highly insulinogenic proteins such as whey are probably beneficial in this regard (29).

This research is intriguing as it may allow athletes to achieve sufficient glycogen resynthesis without having to over-consume carbohydrates. This would be especially beneficial for athletes who need to ensure adequate recovery and glycogen resynthesis between training sessions while keeping caloric intake under control. This would include athletes who are dieting, lighter (female or smaller male) athletes or weight class athletes who need to restrict calories while maintaining training intensity.

Additionally, as I'm going to discuss shortly, the combination of protein and carbohydrate following training has enough other benefits to recommend it beyond simply an effect on glycogen synthesis.

Carbohydrate and protein post-workout: Effects on protein synthesis

As I discussed back in Chapter 3, following training, there is generally an increase in both protein synthesis and breakdown with breakdown exceeding synthesis; thus the body is in a net catabolic state (35). If no nutrients are provided, the body will remain in this state. However, if nutrients are provided following training, the body can be shifted back into a net anabolic state, synthesizing new proteins. I'd mention again that, while both strength/power and endurance training stimulate the synthesis of new proteins following training, the specific proteins synthesized are different.

I'd also note that, while I'm going to focus here on post-workout nutrition, nutrients consumed before or during workout might still be providing nutrients in the immediate post-workout period. That is, a large amount of protein consumed prior to an hour long strength/power workout will still be releasing amino acids to the body following training. However, as the majority of research to date has focused on nutrients consumed after training, that will be the focus of my discussion.

Back in Chapter 3, I discussed the role of insulin and amino acids and their overall effect on protein synthesis and breakdown in terms of how the body responded to meal feeding. The effect of nutrient intake following resistance training is essentially identical: increases in blood amino acid levels have a primary role in stimulating protein synthesis with little impact on protein breakdown while increasing insulin primarily inhibits protein breakdown with little effect on protein synthesis.

As with meal feeding, the essential amino acids (EAAs) and branched chain amino acids (BCAAs), and leucine specifically, appear to play a primary role in promoting protein synthesis following training; the inessential amino acids are not required.

Regarding leucine, a recent study gave protein and carbohydrate or protein, carbohydrate and free leucine following exercise (36). It found higher insulin levels in the free leucine group and a very slightly larger increase in protein synthesis. However, the amount of protein given in the drinks was fairly small (approximately 13 grams per hour for a 75 kg individual along with 6 grams of free leucine), and it's impossible to know whether free leucine would have any further advantage if the protein intake had been higher in the first place. Additionally, the drinks were given for six hours following training (which has little relevance to a single drink taken after a workout) and the overall measured benefit was fairly small.

All high quality proteins contain significant amounts of BCAAs of which some portion is leucine; for example, 40 grams of whey protein contains about 10 grams of BCAAs and 4-5 grams of leucine. It seems unlikely that adding a few more grams of leucine to such an amount of protein would impact greatly on post-workout protein synthesis. Athletes who wanted to reduce the amount of protein consumed post-training for some reason might want to consider adding free leucine to increase the insulin and anabolic response.

As discussed in the last chapter, extracellular concentrations of AAs (rather than intracellular levels) are the driver for protein synthesis and there is a plateau that occurs in terms of how much protein is necessary to generate a maximal anabolic response. AA concentrations above a certain point do not further stimulate protein synthesis (37).

Muscle responds to small amounts of EAAs (3-6 grams EAAs equivalent to 6-15 grams of whole protein) given an hour apart following training (38) but this amount is unlikely to generate a maximum response in the first place. Other research, looking at orally ingested AAs, found no further increase in protein synthesis for 40 versus 20 grams of EAAs (equivalent to roughly 80 or 40 grams of a whole protein) given after training (39). That is to say, both amounts of EAAs generated the same protein synthetic response following training.

To put this in units consistent with the other studies I'm going to discuss, based on the body weight of the subjects, this amounts to roughly 0.15-0.30 g/kg of EAAs, or approximately 0.3-0.75 g/kg of whole protein (30-75 grams protein for a 100kg athlete). Consuming more protein than this post-workout would appear to have no further benefit or effects as the skeletal muscle machinery simply can't utilize all of the AAs past a certain point.

Raising insulin following a workout will primarily be a function of carbohydrate intake. As mentioned above, adding protein to carbohydrate increases the insulin response as well. Consuming high glycemic index carbohydrates following training, along with protein, should provide an optimal insulin response to maximize net protein gain after training.

What the above data all points to is that, following resistance training, protein alone is better than carbohydrate alone but protein with carbohydrate yields superior results to either (40,41). A variety of studies bear this idea out with the combination of carbohydrate and protein having a greater impact on protein synthesis following training than either alone (42,43); carbohydrate by itself is significantly less valuable (44,45).

Although most of the research to date has been done on resistance training, the combination of protein and carbohydrates has also been shown to benefit endurance athletes following training. Providing a combination of carbohydrates and protein after endurance training improves protein synthesis (46) and a recent study found that a protein, carbohydrate and antioxidant drink consumed after a cross-country race significantly decreased markers of protein breakdown and soreness (47).

Endurance athletes seeking maximal performance and adaptation should ensure sufficient intake of both carbohydrate and protein after training. This not only has the potential to improve glycogen resynthesis, but should also improve recovery and overall adaptation to training.

Putting it all together: Amount, timing and type

It's clear from the research that the combination of carbohydrates and protein following training positively impacts on protein synthesis and adaptation in addition to potentially improving glycogen resynthesis following training. Additionally, the consumption of protein and carbohydrates post training appears to impact favorably on anabolic hormone levels following resistance training (40).

Clearly, optimizing the post-workout recovery period means providing AAs in addition to raising insulin by also consuming carbohydrate (48). Towards this goal, it's been suggested that the combination of a protein hydrolysate such as whey (with or without the addition of free leucine) along with rapidly digesting carbohydrates (such as dextrose or maltoxdextrin) might be an optimal combination (48,49). This approach has not been tested in any long-term studies to see if it would provide greater gains compared to say, a protein isolate such as whey or milk protein isolate.

Given the extremely slight differences in absorption time for casein and whey and their hydrolysates (discussed in Chapter 2), I find it unlikely that major differences would be seen using a hydrolysate compared to a protein isolate. This is even more likely to be true if my recommendations for pre- and during-workout nutrition are being followed.

With those "meals" still providing AAs immediately after training, there would seem to be no real priority in terms of providing AAs as rapidly as possible with the use of a hydrolysate. As discussed above, given sufficient amounts of high quality protein in the first place, I tend to doubt that adding extra free leucine will have much of an additional impact.

Methodological problems with the studies: Fasted vs. non-fasted subjects

Before I continue with this chapter, I want to discuss one methodological issue relating to the studies done on nutrient timing and resistance training. With very few exceptions, most of the work has used fasted subjects. That is, the subjects come in to the lab after an overnight fast, train, and then various nutrient combinations are consumed and the impact on protein synthesis or breakdown is measured.

While this makes control of the study variables simpler, it does raise some questions regarding the applicability of some of the research. For example, consider the study mentioned above that found that consuming EAAs before training yielded better results than consuming them afterwards (8).

In an overnight fasted state, clearly a pre-workout meal is going to provide nutrients at a sooner time than anything consumed after training. However, what if the subjects had had a solid meal one to two hours before training and thus had protein and carbohydrates digesting and being released into the bloodstream? Would pre-workout nutrition still have yielded superior results?

About the only study that has examined pre/post workout nutrition under more realistic conditions is the study by Cribb mentioned above and discussed in more detail below (11). As described above, it compared a group that consumed nutrients immediately before and after training (but within the context of a normal day of eating) to a group consuming the same nutrients at other times of the day and found that the pre/post workout group got better results in terms of LBM gains and fat loss.

It's also unknown at this point whether combining the results from all of the individual studies will yield greater results. For example, as discussed above, consuming nutrients both before and during resistance training has been shown to have benefits in terms of promoting protein synthesis and limiting protein breakdown. As I discuss below, consuming nutrients afterwards does as well.

Does that mean that consuming nutrients before and after is better than only one or the other? What if you consume nutrients before, during and after; is that better than just before, or before and after, or just during?

As described in the sections above, pre-, during- and post-workout nutrition each have slightly different benefits and effects in terms of exercise performance, recovery and adaptation. Based on physiological and theoretical concerns, there is some logical reason to believe that optimizing nutrition around training by providing nutrients through all three time periods (pre-, during- and post-workout) should yield better results than only consuming nutrients at one or two of those times. This idea has not been systematically tested.

An additional issue regarding most of the research to date, which I'll address below, is that the grand majority of studies have been short-term, examining only the acute response to nutrient ingestion after training. Only a handful of studies have examined whether or not consuming nutrients around training yields improved gains in lean body mass, strength or performance in the long-term.

Timing

Based on early data looking at glycogen resynthesis, the idea of a "window of opportunity" in terms of post-workout nutrient intake has long been discussed. For example, early research found that the consumption of carbohydrate/protein 1 hour following endurance training yields better results than waiting until 3 hours after training (50). It's less clear whether the immediate consumption of protein/carbohydrate after resistance training is required for optimal gains.

In elderly individuals, the consumption of protein/carbs immediately after training, as opposed to waiting two hours, significantly improved muscle gain (51) while in younger individuals whether aminos/carbs were given one or three hours following resistance training, the effects on acute protein synthesis were the same (52). Older individuals typically respond differently than younger individuals (consider the difference in response to protein "pulse" feeding between older and younger subjects discussed in Chapter 7); this may explain the difference in results between these two studies.

Additionally, if nutrients were consumed before and/or during training, amino acids and carbohydrates will still be available in the post-workout period. Whether consuming additional nutrients immediately after training would be of additional benefit is currently unknown.

As mentioned above, from a physiological standpoint, it makes logical sense to provide nutrients immediately after training to shift the body back into an anabolic state. In addition, blood flow to the trained muscles is typically higher following training, as is AA uptake (53). Providing additional nutrients at that time would certainly seem to be beneficial.

However, if pre- and/or during-workout nutrition is consumed, it may not be crucial to consume a post-workout shake or meal immediately after training as breaks of one to three hours appear to have no major impact in the short-term.

Type and amount

Because of the way the studies have been performed, it's impossible to separate the type of protein from the amount of protein given so I'm going to discuss them together. In terms of resistance training, I'm not aware of any research systematically examining different carbohydrate types; most studies have typically used sucrose (table sugar) although the Cribb study mentioned above used dextrose before and after training.

Early studies used a combination of EAAs, generally with sucrose in varying amounts and timing. One study found that the consumption of 3-6 grams of EAAs one and two hours after training stimulated protein synthesis each time (38) but as mentioned above, this is unlikely to have maximized the anabolic response in the first place. Given the research discussed in Chapter 7 on skeletal muscle becoming "desensitized" to AAs, had more protein been given at the one-hour mark, it seems unlikely that the second dose would have had an effect.

Tipton provided 6 grams of EAAs with 35 grams of sucrose either before or after training (8). The pre-workout EAAs generated a larger impact on post-workout protein synthesis than the same dose given post-workout. As mentioned above, this appears to mainly have been due to one subject who had a massive response to the pre-workout nutrients. As well, the study was done fasted which raises some issues about the relevance of the findings to most real-world eating patterns (as discussed above).

Another study determined that a maximum response in terms of protein synthesis following training was found with 20 grams of EAAs after training (39). This would equate to roughly 40-50 grams of whole protein; based on the body weight of the subjects this amounts to approximately 0.6 g/kg of whole protein to generate a maximal anabolic response after training.

While a majority of the early research tended to use EAA supplements around training, more recent research has begun to examine the impact of whole proteins on protein synthesis post-training. The Cribb study described above gave bodybuilders 0.40 g/kg (0.18 g/lb) of whey protein, 0.43 g/kg (0.2 g/lb) of dextrose and 0.07 g/kg (~0.03 g/lb) of creatine with trace fat before and after training. For a 100kg (220 lb) lifter this provided 40 grams of whey with 43 grams dextrose and 7 grams of creatine (11).

Borsheim provided 17.5 grams of whey protein (0.23 g/kg) with 77 grams carbohydrates (1.0 g/kg) and 5 grams of amino acids following resistance training (45), not unsurprisingly this yielded superior results to 100 grams of carbohydrates consumed alone.

Another study examined the impact of either 20 grams (~0.26 g/kg) of casein or whey after training; despite differences in the pattern of amino acid appearance in the bloodstream, both produced an identical result in terms of protein synthesis (54). Readers may recall from an earlier chapter that a mixture of whey and casein has been hypothesized to be superior to either alone in the post-workout period (55). Recent work suggests that this is likely to be the case.

One recent study examined the impact of skim milk, whole milk, or enough skim milk to match the caloric content of the whole milk on protein synthesis following resistance training. This provided either 8 or 14 grams (0.13-0.19 g/kg) of protein along with 12 or 20 grams of carbs (56). Interestingly, whole milk yielded better results than skim milk although the reasons for this are unclear.

In another study, 18 grams (0.22 g/kg) of skim milk protein (with 24 grams carbs and 1.5 grams fat) was shown to generate a superior anabolic response to a soy based drink containing identical nutrients and calories following resistance training (57). Soy is a fast protein, similar to whey, and the difference in response appears to be related to the slower digestion of the milk protein. The same group found a trend towards increased lean body mass gains with milk compared to soy protein in longer- term studies as well (58,59).

Similarly, a recent study compared the intake of 48 grams of carbohydrate (0.57 g/kg) to 40 grams whey with 8 grams casein (0.57 g/kg total protein) or 40 grams of whey with 5 grams BCAAs and 3 grams of leucine over 10 weeks of resistance training. The casein/whey mixture yielded the best gains in muscle mass (60). However, differences in caloric intake (e.g. the whey/BCAA/glutamine group ate fewer calories) color this observation.

Collectively, the above research seems to suggest that consuming slower digesting proteins, or a mixture of fast and slow proteins during the post-training period may provide superior results compared to a fast protein. Although no study has tested it directly, milk protein isolate (MPI), which contains 20% whey and 80% casein might turn out to be an ideal post-workout protein; the studies using milk (skim or whole) would certainly seem to support this idea.

By providing a rapidly available source of amino acids (from the whey) with a slower digesting source (from the casein), both protein synthesis and breakdown should be positively affected; this effect should be even greater when combined with carbohydrates to raise insulin.

Research on endurance training has mostly focused on glycogen resynthesis more than protein synthesis or recovery. I'm not aware of any research systematically examining the impact of different amounts or types of protein following endurance training on either glycogen resynthesis or adaptation. Looking at amounts, one study found that the addition of 40 grams of protein to 110 grams of carbohydrate increased glycogen resynthesis following training (32). Similarly, 28 grams of protein added to 80 grams of carbohydrate resulted in greater glycogen resynthesis compared to either 80 or 108 grams of carbohydrate (61). As mentioned above, any benefit of adding protein to carbohydrate in terms of glycogen synthesis appears to be lost if sufficient carbohydrate (1.5 g/kg or more) is provided to begin with (24) or the carbohydrates are given at frequent intervals.

Examining the impact of a protein/carbohydrate drink on muscle damage following distance running, one study provided roughly 1.5 g/kg of carbohydrate (~0.7 g/lb) with 0.35 g/kg (0.15 g/lb) of protein following training (47). For a 60 kg (132 lb) athlete this provided 88 grams of carbohydrate and 21 grams of protein. The subjects in the protein/carbohydrate group reported less muscle soreness. Markers of muscle damage were also reduced following training; in the long-term this might be expected to improve performance.

Another study compared chocolate milk to a carbohydrate only drink consumed between two endurance sessions set only 4 hours apart. Subjects received 0.27-0.42 g/kg protein and 1-1.2 g/kg carbohydrate along with a small amount of fat from chocolate milk (roughly 2-3 cups of chocolate milk) or one of two carbohydrate only drinks (62). The chocolate milk group performed better at the second session indicating that the combination of carbohydrate and protein was superior for recovery between workouts.

Although unrelated to recovery per se, milk was found to be superior to either water or Powerade (a commercial carbohydrate drink) for rehydration following endurance exercise, most likely due to the higher sodium and potassium content (63).

Overall, as with resistance training, it appears that consuming protein and carbohydrates after endurance training is beneficial, both from a glycogen resynthesis and recovery point of view. There is some indication that the practice of proper post-workout nutrition can also limit muscle damage while promoting positive adaptations to training.

A final issue that has received relatively little study is whether liquids or solids post-workout are superior. From a glycogen resynthesis point of view, it doesn't appear to matter whether liquids or solids are consumed following training (64).

In terms of resistance training, I'm only aware of one study which compared a solid and liquid meal following resistance training and examined a variety of hormonal parameters (65); no measures of protein synthesis or breakdown were made. The meal provided 38 grams protein (0.41 g/kg), 70 grams carbohydrate (0.75 g/kg) and 7 grams fat. The liquid meal was a blend of whey, casein and milk protein along with maltodextrin, sucrose and dextrose with canola, MCT and borage providing the fat; the solid meal consisted of skinless chicken breast with rice.

The only difference seen was that the liquid meal maintained insulin at a higher level compared to the whole food meal, possibly due to the insulinogenic effects of whey or the faster rate of digestion for liquids versus solids. This could potentially stimulate protein synthesis (or inhibit protein breakdown) more effectively although this was not examined in the study.

Of practical importance, the subjects reported finding it easier to consume the liquid meal following training. This is not uncommon as many athletes find their hunger blunted following intense training. Practically speaking, a liquid meal often makes it easier to consume sufficient post-workout nutrients.

As well, athletes looking to maximize glycogen replenishment for a second training session later in the day would likely benefit from the increased insulin response; they would also be able to eat again sooner by not having a solid food meal digesting. Finally, liquid nutrients following training would also go towards rehydration following the workout.

My recommendations for post-workout nutrition appear in Table 4 below. Continuing with the theme of this chapter, endurance athletes will be consuming more carbohydrate and relatively less protein following training while a strength/power athlete would be consuming relatively less carbohydrate and more protein.

Table 4: Post workout nutrition recommendations

	Protein	**Carbohydrate**
Endurance	0.15-0.35 g/kg	1-1.85 g/kg
Strength/power	0.3-0.5 g/kg	0.3-1.5 g/kg

The ideal protein post-workout is unknown although emerging data suggests that a slower protein such as casein or a mixture of fast/slow proteins such as MPI (or a half and half mixture of casein and whey) might be ideal. Carbohydrate sources ranging from dextrose to maltodextrin (a long chain polymer of glucose) with some fructose (to refill liver glycogen, important for overall anabolism) can be consumed. Waxy maize starch (commercially sold as Vitargo) may be beneficial for athletes who need to maximize glycogen synthesis between two workouts which are very close together although I tend to doubt it has much benefit for athletes training only once per day (where maximum rates of glycogen resynthesis are not required).

Athletes who prefer to consume whole food following training should generally pick somewhat more easily digested foods such as lower fat protein (chicken, fish, low fat red meat) along with higher GI carbohydrates (potato, breads, rice, etc.).

The range for carbohydrate intake post-workout is necessarily large due to the differences in the types of workouts that might be included in the above categories. The values above are given in an attempt to cover the myriad possibilities that might come up and I want to discuss some of those eventualities here (I'll address this issue in more detail in Chapter 13).

Following endurance workouts, carbohydrate amounts would be chosen depending on the extent of the workout (longer and/or more intensive workouts would require more carbohydrate) along with how rapidly the athlete needed to replenish glycogen. An athlete training once per day could go with the lower amounts of carbohydrate so long as sufficient carbohydrate is consumed over the next 24 hours. An athlete who needed to train again later that same day would want to consume larger amounts of carbohydrates to maximize the immediate resynthesis of glycogen.

How much protein is consumed will tend to depend on the size of the athlete and total amount of protein being consumed per day. A smaller female endurance athlete consuming the low end of my daily protein recommendations (1.3 g/kg) may only consume 0.15 g/kg of protein post-workout to ensure that sufficient amounts can be consumed at other meals while a larger male athlete consuming the high-end recommendations of 2.0 g/kg may go with the larger amount of protein following training.

Workout intensity or duration plays a role here as well, relatively longer and/or more intensive workouts would tend to require larger amounts of protein following training than shorter and/or lower intensity workouts. Running, because of its higher impact nature (compared to other endurance sports) tends to cause more muscle damage as well, proportionally more protein might be needed after running workouts compared to sports such as cycling, swimming, rowing, etc. This has not been studied to my knowledge.

For strength/power workouts, the choice of post-workout carbohydrate amount depends on the nature of the workout (volume and type of training) along with the goals of the workout. The larger the volume of work in the glycolytic range (work bouts ranging from roughly 30-90 seconds), the more glycogen that is used and the more it will be necessary to replenish glycogen following training. An athlete performing a large amount of bodybuilding type of work would require more carbohydrates following training than an athlete performing lower repetition "neural" training.

Additionally, an athlete looking for maximal gains in body mass would tend towards the higher amounts of both protein and carbohydrate after training when compared to an athlete who was simply trying to maintain body weight (or explicitly avoid mass gains); the latter athlete would consume relatively less carbohydrate (and possibly less total protein as well).

Athletes performing other types of strength/power workouts such as sprinting or plyometrics will have to use a bit of trial and error to determine how many carbohydrates they need following training. In general, the shorter the work bouts (less than 30 seconds), the less glycogen that will be depleted; hence less carbohydrates will be necessary. As above, the more work done in the time range of 30-90 seconds, the more glycogen that will be used and the more carbohydrates should be consumed following training.

Acute versus long-term studies

As mentioned above, one limitation with many of the post-workout studies is that they generally only examine what happens immediately after training. A question unanswered until fairly recently is whether these acute changes in protein synthesis and breakdown actually result in a greater amount of muscle gained over time.

While this may seem like a silly question (logically, how could promoting protein synthesis after a workout not generate better gains?), it's actually not that simple. It's always possible that any adaptation seen in the short-term could be negated by a counter-adaptation later in the day. For example, perhaps the small gain in protein following training plus nutrient ingestion is counterbalanced by a greater breakdown of protein later in the day (as occurs when non-training subjects simply increase their protein intake).

However, more recently, the answer to this question has become much more clear. One early study showed that the protein gained after training with post-workout nutrients resulted in a greater 24-hour protein balance (66). That is, the body simply added the protein gained from training plus nutrients to its total stores. However, this still didn't address whether lean body mass gains would be improved over time. Additionally, since there was no control group that trained without consuming nutrients after workout, it was impossible to know if the results were from the training, the protein supplement or the combination of the two.

Earlier, I mentioned a study which compared carbohydrate only to whey/casein or whey/BCAA/glutamine over 10 weeks and found that the whey/casein group showed greater gains in lean body mass compared to either other group (60). However, there was no placebo group consuming nothing after training for comparison. Perhaps the subjects gained more lean body mass as a consequence of consuming more calories with no impact of the actual timing of consumption

One recent study compared either carbohydrate alone to protein consumed before and after training over 14 weeks of resistance training (67). In that study, only the protein group showed gains in muscle size but, again, with no control group, it's impossible to conclude that consuming nutrients around training was the cause. Perhaps simply consuming more protein per se, rather than the timing, was responsible.

Although verification is necessary, the recent study by Cribb at least points towards the logical conclusion being correct (11); consuming nutrients around training (when utilization and uptake is greatest) leads to greater lean body mass gains compared to consuming an identical amount of nutrients at other times of the day. I'd refer readers to the discussion above for more details on this paper.

I am unaware of any research examining whether pre-, during- or post-workout nutrition will impact on the long-term adaptations to endurance training although there are certain theoretical reasons to believe that this would be the case. Similar physiological effects in terms of blood flow, amino acid uptake and protein synthesis are occurring following endurance training. It makes sense that shifting the body back into an anabolic state by consuming nutrients following all types of training *should* serve to promote better long-term adaptations.

Remaining questions about optimal nutrition around workouts

Research to date has examined a number of different aspects regarding the impact of pre-, during and post-workout nutrition but questions still remain as most of the work has been done in isolation and many different intake patterns are possible.

If an athlete consumes nutrients immediately pre-workout, is post-workout nutrition absolutely required? What about consuming nutrients during training following pre-workout intake? If an athlete consumes nutrients immediately before and during training, can they simply wait an hour to eat a solid meal or will extra nutrients consumed immediately after training yield better results?

If an athlete has consumed a solid meal two hours prior to training, we might ask if any of this is really necessary in the first place. As discussed in Chapter 7, solid food meals take hours to digest, meal consumed within two to three hours of training will still be digesting and releasing nutrients to the body throughout the workout period; does this make around workout nutrition moot?

Assuming that consuming nutrients around training will still yield better results given a normal meal schedule, the unanswered questions change. Is it better to spread nutrition consumed around training evenly across the pre-, during and post-workout period or will some other pattern turn out to be optimal? At this point, there are more questions than answers.

Of the research examined above, only a few studies have even partially examined this issue and neither was making a comparison of different patterns of consumption. As mentioned, the study by Cribb compared pre- and post-workout nutrition to the same nutrients consumed at times other than around training; the pre/post group showed better gains in lean body mass with a slight fat loss (11).

A recent study on resistance training provided a small amount of carbohydrate and protein (62 and 15 grams respectively) starting 30 minutes before to immediately after training, this decreased markers of protein breakdown but, as the study only looked at a single workout, the impact over the long-term is unknown (21). However, the study wasn't comparing various intake patterns to one another; rather, it simply spread the nutrients around the workout period.

A very recent study examined the impact of resistance training on molecular markers of hypertrophy (68). Trainees were given either 0.3 g/kg of protein or 0.3 g/kg protein with an additional 0.3 g/kg of carbohydrate before, immediately after, and again 1 hour after training (so they received a total of 0.9 g/kg of carbs and protein spread around training). The group receiving carbohydrate and protein showed the greatest impact on the markers examined. Once again, the study didn't examine different intake patterns; it simply provided nutrients throughout the immediate pre- and post-workout period to ensure a maximal response.

Given that each period of around training nutrition can be seen as having a slightly different impact on training, recovery or adaptation, my general feeling is that spreading nutrients throughout the pre-, during- and post-workout nutrition should yield optimal results.

Providing carbohydrate and protein before workout ensures optimal blood glucose and AA levels during training. During workout nutrition appears to limit protein breakdown and fatigue during both endurance and resistance training and may improve overall workout performance in some situations. Post-workout nutrition optimizes recovery in terms of both glycogen and protein synthesis; this improves long-term adaptations to both types of training.

At the same time, there are some practical reasons why athletes might choose to forego some aspect of around workout nutrition. Some athletes don't like training with anything in their stomach, or find that their blood sugar levels get unstable with pre- or during-workout carbohydrates. This can be especially true for sprint and low-repetition weight workouts where the rest-intervals are often long; it tends to be much less of an issue for endurance athletes. This problem can be compounded if very fast acting carbohydrates like dextrose are chosen; sucrose or Gatorade type drinks tend to avoid problems with blood sugar crashing, probably because the fructose helps to maintain blood sugar.

Assuming that a sufficient solid meal is consumed within 2-3 hours of training, such that nutrients are still being released during the workout, athletes may find that they can eliminate the pre- and during-workout phases and only consume post-workout nutrition. Or they might skip the immediately pre-workout nutrition and only consume a small amount of carbs/protein during the workout to maintain blood glucose and AA levels, followed by a post-workout drink (or solid meal). As mentioned above, outside of some practical considerations, athletes who only train once per day or less may prefer to consume a solid meal following training and avoid liquid nutrition altogether.

I'd also mention that some of the choice of around workout nutrition will also clearly depend on the length and type of workout. A bodybuilder performing an hour of lifting will have far less requirements for nutrients in terms of maintaining blood glucose or amino acid levels than a sprinter or athlete performing a workout lasting two or more hours. A single meal consumed two to three hours before training would probably be insufficient in that situation. The same would hold for an endurance athlete who may be performing a several hour training session, simply eating a few hours before training would be unlikely to keep them fueled during training.

Nutrient timing and body fat

One potential concern that is often raised regarding nutrition around training is the potential impact on body fat levels; this tends to be especially true in the bodybuilding community but is also relevant to athletes who need to maintain a very low level of body fat for optimal performance.

The concern typically revolves around the consumption of large amounts of high glycemic index carbohydrates and raising insulin around training; a number of creative post-workout nutrition strategies have been developed in an attempt to promote recovery while minimizing an increase in insulin. Is this concern valid?

Research has repeatedly shown that muscular insulin sensitivity is up regulated following both resistance and endurance training although the mechanisms are different for each type of training (69). Improvements in insulin sensitivity also depend on the total amount of training done with a threshold caloric expenditure of 500-900 calories per workout being necessary (70,71). While these levels may exceed what a typical sedentary individual can accomplish, even moderately well trained athletes should be able to easily achieve them.

The training induced increase in insulin sensitivity is likely to be a major mechanism behind the "window of opportunity" following training, skeletal muscle is primed to take up nutrients at an accelerated rate. This means that the chance of calories being pushed into fat storage is minimized if not eliminated. Related to this, following glycogen depleting endurance exercise, research has clearly shown that the ingested carbohydrates go towards glycogen storage while the body continues to rely on fatty acids for fuel (72, 73). This occurs despite an increase in insulin levels from the carbohydrates.

This holds true for massive carbohydrate intakes as well. In one study, subjects were given 5 g/kg of carbohydrate (500 grams carbohydrate for a 100kg athlete) following 90 minutes of moderate exercise and de novo lipogenesis (DNL, the conversion of carbohydrate to fat) was studied; not only did no DNL occur but the body continued to burn fat in the post-exercise period (74). Essentially, when glycogen is depleted from training, incoming carbohydrates go to glycogen storage while the body continues to use fatty acids for fuel; raising insulin post-workout does not interfere with this.

Related to the above, various concerns regarding the hormonal response to training (especially in terms of testosterone and growth hormone) have occasionally been raised. The debate over the potential relevance of the hormonal response to training is beyond the scope of this book; suffice to say it's unknown how important fairly small and acute (one hour or less) changes in hormones are to the overall adaptation to training. However, even if we assume that they are important, it doesn't appear that consuming nutrients around training negatively impacts upon them.

One of the early post-workout nutrition studies found that carbs and protein consumed after resistance training led to a more pronounced growth hormone (GH) response than protein alone or placebo (40). A later study showed that consumption of 50 grams of carbs and 25 grams of protein before and after training led to an increased growth hormone (GH) response (75). Interestingly, that same study showed that post-workout feeding caused a lowering of testosterone in the post-workout period which might be interpreted as a negative from the standpoint of muscle gain. However, this decrease is thought to represent increased clearance of testosterone into skeletal muscle rather than a decrease in production per se.

In support of this idea, a recent study showed that consuming carbohydrate, protein and fat after training led to an increased level of androgen receptors (the receptor that testosterone binds to in order to stimulate gains in skeletal muscle) compared to water alone (76). Increased androgen receptor levels would bind more testosterone, possibly explaining the decrease in blood levels following training. Other research, already discussed, has shown that consuming nutrients during or after workout leads to increased insulin and decreased levels of cortisol, indicating a more anabolic state.

Related to this issue, a question that often comes up is whether carbohydrates and calories should be consumed around or during training when an athlete is trying to lose body fat. Based on the data above, it seems unnecessary to limit calories around training. Readers might consider again the study by Cribb which found a slight fat loss (1 kg) in the group consuming dextrose, whey protein and creatine before and after training (11).

Quite in fact, given the crucial importance of ensuring workout intensity as well as promoting recovery during dieting, it seems backwards to limit calories around training. Instead, caloric reductions should be made at other meals of the day with appropriate pre-, during- and post-workout nutrition being maintained. This ensures proper recovery while dieting while still allowing for a suitable deficit to be created. Athletes having to restrict calories to reduce body mass may wish to reduce their around workout nutrition to the lower values in the recommendations, this will allow more food to be consumed at the other meals of the day. This topic is addressed in further detail in Chapter 13.

Application

Although I'll discuss this in more detail in Chapter 13, I want to talk about how all of the information in this chapter can be put into practice.

The simplest approach, of course would be to simply take all of the values given above for pre-, during- and post-workout nutrition and total them up. That amount of carbohydrate and protein would be mixed into a sufficient amount of liquid in a single large bottle. The athlete would simply begin sipping the drink starting 30 minutes before the workout, continuing through the workout and finishing the drink at the end of the workout.

Let's say we have a 100 kg (220 lb) strength/power athlete who will consume 0.5 g/kg protein before and after training for a total of 1.0 g/kg (100 grams of protein) along with an additional 15 grams of protein during the workout. So 115 grams of protein total around training. To that he might add an equal amount of rapidly digested carbohydrate or 115 grams. Creatine might also be added to the drink which would be mixed up in roughly 64 oz of fluid. This would be sipped starting 30 minutes before workout, consumed evenly throughout the workout and finished when the workout was over. A solid meal would be consumed 2-3 hours later and normal eating resumed.

The drawback to the simple approach, of course, is that mixing everything into one bottle mandates that the same protein and carbohydrates be consumed throughout the entire around-workout nutrition period. This may not be optimal.

As discussed in detail in this chapter, there is some indication that different proteins (and possibly carbohydrates) might be superior at different times. While fast proteins are probably optimal before and during a workout, a slow or slow/fast mixture appears to be ideal following training. Additionally, while fast acting carbs such as dextrose are most appropriate after a workout, they can cause blood sugar problems in some athletes. A mixture of glucose and fructose, or sucrose, might be a better choice for consumption during training under those conditions.

A slightly more complex approach would be to mix two drinks, one to be consumed pre- and during workout (since whey protein can be used during both time periods) with a second bottle (containing a slow or mixed protein such as a casein/whey mixture or milk protein isolate) mixed for post-workout. Different carbohydrates could be used in each drink with this approach.

Obviously the most time intensive approach would be to mix three drinks; this would allow the greatest amount of control over the nutrient (and liquid content) of each phase for athletes trying to optimize all aspects of their around training nutrition.

For example, an athlete might choose to consume 25 grams of dextrose with 25 grams of whey protein approximately 30 minutes before training in only a small amount of fluid, to avoid issues of gastric upset (or having to pee during training). This should give the whey sufficient time to digest to provide AAs to the working muscles.

During the workout, an additional 60 grams of a typical hydration drink (such as Gatorade) with 12-15 grams of whey protein might be consumed in 32-36 oz of fluid every hour. This would maintain blood glucose and blood amino acids along with ensuring sufficient fluid intake.

Post-workout, a mixture of dextrose (or maltodextrin, or a mixture of the two) and milk protein isolate (or a half and half mixture of whey and casein) could be consumed in however much water was necessary to ensure optimal rehydration following training. The amounts used would, of course, depend on the athlete's goals, the amount of training that had been done and the other issues discussed throughout this chapter.

9

Protein Controversies

Before looking at whole proteins and protein powders, I'd like to address some of the most common controversies that tend to surround the high protein intakes typically seen in and recommended to athletes. The major ones are kidney function, bone health, and heart disease and colon cancer. Related to the issue of bone health, I'm also going to address the topic of metabolic acidosis and the impact that dietary protein intake has upon it.

Kidney function

A common criticism of high protein intakes/diets is the concern that they are damaging to the kidneys. This belief seems to stem from the fact that, in individuals with preexisting kidney damage, protein intake often has to be reduced to prevent further development of the disease. Incorrectly, this has been turned around to suggest that high-protein intakes are damaging to the kidneys (1).

There is at best a weak case to be made for a risk of high protein intakes on kidney function; quite in fact, some research suggests a *beneficial* effect of higher protein intakes on kidney function (2). Simply put, the adaptations to kidney function that are often cited as indicating "strain" or damage are more likely to simply be normal adaptive effects of varying protein intake (1).

Unfortunately, very little research has directly examined the impact of high protein intakes on kidney function in athletes. One study examined the impact of 2.8 g/kg (1.3 g/lb) protein on the kidney function of bodybuilders, no negative effect was seen (3). To my knowledge, higher intakes have not been studied.

Empirically, it's worth considering that athletes have been habitually consuming large amounts of protein for at least several decades without any reported increase in the incidence of kidney problems. If such a problem were going to occur, it seems likely that it would have shown up by now. While this certainly doesn't prove that high protein intakes aren't potentially detrimental to kidney function, the data in support of that idea would seem to be lacking both from a scientific and real-world point of view.

Interestingly, while it's always been stated that high dietary protein intakes increase fluid requirements, this idea appears to have originated from a military study examining nitrogen balance under conditions of water and energy restriction (1). There is no indication that individuals who are sufficiently hydrated need to go out of their way to increase fluid intake when they are consuming large amounts of protein.

Bone health

Perhaps one of the most pervasive criticisms of high protein intakes has to do with the impact of protein on bone health and calcium status. This goes back to early nutritional studies which gave purified protein diets and saw a loss of calcium from the body.

Later studies, using whole food proteins (which included other nutrients such as phosphorous) found very different effects. Frankly, the early studies on this topic are fairly irrelevant to normal human nutrition since the consumption of protein in the total absence of other nutrients would be extremely rare; all whole food proteins and protein powders contain micronutrients.

As well, the impact of protein on overall calcium status is more complex than having a simple positive or negative effect as dietary protein can impact on both calcium excretion as well as calcium absorption and utilization. It is the combined effect of these processes which determines the end result in terms of bone health.

In epidemiological studies, a high intake of animal protein increases the risk of bone fractures; as well, a high ratio of animal to vegetable protein intake has also been associated with an increased risk of bone loss (4). In contrast, high intakes of protein improve bone healing, following a fracture for example. This is mediated both by increased calcium absorption as well increased levels of insulin-like growth factor 1(IGF-1), a hormone involved in tissue growth (5). How can this contradiction be reconciled?

Fundamentally, it's too simplistic to look at protein intake in isolation in terms of its effects on bone health as the protein content of food interacts with other nutrients in that food or in the total diet (6). For example, recent studies suggest an interaction between protein and calcium intake.

When calcium intake is low, high protein intakes appear to have negative effects on bone health. In contrast, when calcium and vitamin D intake are sufficient, protein intake has a beneficial effect on bone health (7). This suggests that ensuring adequate calcium intake (through a sufficient intake of dairy foods, or calcium supplements) is crucial for bone health when large amounts of protein are being consumed.

This most likely serves to explain the above contradiction. In the studies where dietary protein intake was found to have a negative impact on bone health, there were other dietary factors playing a role. Calcium or Vitamin D intake may have been insufficient causing an overall negative effect. However, when sufficient calcium and Vitamin D are provided (as they typically are following bone injury), dietary protein has a beneficial impact.

Metabolic acidosis

Related to the issue of dietary protein and bone health is a concept referred to as net renal acid load (NRAL). When foods are consumed, they have the potential to produce either a net acidic or net alkaline (basic) effect, which the body, primarily the kidneys has to deal with. NRAL refers to the total amount of acid produced that the kidneys have to process.

Simplistically, protein foods tend to increase the net renal acid load, as does a high intake of sodium relative to potassium. In contrast, fruits and vegetables, along with foods high in potassium, tend to buffer this net acid load and have an overall alkalizing effect on the body. With an excess of acid forming foods in the diet relative to the number of base producing foods, a slight metabolic acidosis can occur.

The modern diet, with its high reliance on animal proteins and high intake of sodium, along with a low intake of fruits, vegetables and potassium is thought to generate a sub-clinical metabolic acidosis (8). Even a slight increase in the overall acid status of the body can have a number of negative health effects, not the least of which is an impact on hormones important to athletes (9). Ensuring sufficient intake of basic foods (fruits and vegetables) to balance out the acid produced from a high protein intake is one key to avoiding this problem.

From both a bone health and performance standpoint, any athlete consuming a high protein diet must ensure sufficient intake of other foods including plenty of fruits and vegetables to buffer any potential negative effects (10). Using a potassium salt or mixed sodium/potassium salt to ensure adequate potassium intake to offset the high levels of sodium in the modern diet is not a bad idea either.

As a final comment related to this issue, it has been suggested that the impact of diet on the body's acid balance can impact on exercise performance. It's well established that low-carbohydrate diets tend to decrease the body's ability to buffer acid produced during high intensity exercise, for example. This hurts performance during those types of events. Reducing protein intake and increasing carbohydrate intake for 3-5 days prior to an important event has been theorized to increase exercise performance in events lasting 3-7 minutes (11).

Colon cancer/heart disease/overall health

A large meat intake, especially red meat, is often claimed to be involved in the development of a number of diseases, especially heart disease and colon cancer. A great deal of this research is based on observational work where individuals consuming a meat-based diet are more likely to get such diseases. As well, there is ample evidence to suggest health benefits with vegetarian diets (12).

However, as with the protein and bone health issue, you can't simply isolate protein/meat intake from other aspects of the diet. This is important when looking at the research as most of it tends to be epidemiological in nature, that is it looks at large populations of individuals and tries to draws correlations between different measured variables. This can lead researchers to draw incorrect conclusions.

For example, modern meat based diets are also typically very high in fat with typical cuts of red meat being high in saturated fat, a known risk factor for various diseases. In contrast, lean red meats, trimmed of visible fat, have a drastically different impact on the risk of cardiac disease (13). As well, unprocessed lean red meat doesn't increase markers of inflammation or oxidation (14). In addition to potential cancer promoting factors, meat also contains a number of cancer preventing factors (15). Replacement of carbohydrate with lean red meat has also been shown to *lower* of blood pressure (16). The key here, of course, is that *lean* red meat, as opposed to the fattier cuts commonly consumed were studied.

Diets high in meat are often low in fruits and vegetables (meaning a low intake of important micronutrients as well as fiber) and research suggests that it is the lack of those foods (fruits, vegetables) more so than the presence of red meat that is responsible for any increased cancer risk (17). High fat intakes have also been associated with low food variety and low intakes of fruits and vegetables (18); this would further contribute to the apparent link between consuming fatty meat and health risk.

Put differently, there is going to be a fairly large difference in the overall impact of a diet that is high in animal protein, high in fat, low in fruits and vegetables (and thus low in fiber and other important nutrients) which may be accompanied with other health risks such as inactivity, being obese, etc. This would be held in complete contrast to an athletic diet containing large amounts of lean meats along with a large fruit and vegetable intake, high levels of activity, maintenance of a low level of body fat, etc.

As I mentioned above with regards to bone health any diet high in animal protein must be accompanied by a high intake of fruits and vegetables. As well, leaner cuts of meat (especially red meat) should be chosen whenever possible.

Summary

A number of health risks have been attributed to the consumption of high protein intakes, this includes potential problems with the kidneys, bone health, metabolic acidosis and certain types of cancers. For the most part, these risks tend to be extremely overstated.

While high protein intakes may cause problems when there is pre-existing kidney disease, no research suggests that high protein intakes cause kidney damage. While there is potential for high protein intakes to cause body calcium loss, this appears to only occur when calcium intake is insufficient in the first place; high protein intakes with high calcium intakes improves bone health. Ensuring sufficient vegetable intake along with a high protein intake is a key aspect not only to bone health but to preventing a small metabolic acidosis which may occur when large amounts of protein are consumed by themselves.

Concerns over heart disease and cancer are more related to the high fat content of many cuts of meat, along with other nutritional factors such as insufficient fruit and vegetable intake that contributes. Other lifestyle factors that typically accompany the consumption of higher fat cuts of meat are also a likely contributor to the overall health risk. The consumption of lean cuts of meat has actually been shown to improve overall health; both athletic and diets for general health should ideally contain plenty of fruits and vegetables for this reason.

10

Whole Food Proteins

Overall, I feel strongly that an athlete's daily protein intake should come primarily from whole food sources. While supplements offer convenience, and many are now more economical (in terms of cost per gram protein) than food proteins, there are likely to be nutrients present in whole foods that are as of yet undiscovered by research, which are important to either human health or athletic performance.

The proteins discussed in this chapter are not presented in any order of importance and I'll make more specific protein recommendations for different situations in Chapter 13. For the most part, all of the proteins discussed in this chapter are appropriate at all times of the day except immediately before and during training; I'll discuss any exceptions as necessary.

By way of introduction

Although this book's primary focus is the role of protein for athletes, examining various protein sources can go beyond simply looking at the amount or quality of the protein. Whole proteins contain other nutrients or compounds that may offer nutritional benefits (or negatives) and this should play a role in the selection of protein by athletes or the general public.

For example, the fat content of a given protein may be an important consideration for an athlete trying to limit fat intake (or increase it for that matter). Even that is complicated by the type of fat present. Some higher-fat protein sources contain significant amounts of healthy fats such as the omega-3 (w-3) fish oils; others contain trace amounts of fats such as conjugated linoleic acid that may play a functional role in the human body. An additional consideration would include the presence or absence of micronutrients such as zinc, iron, B12 and calcium.

Of course protein foods may also have potential health negatives associated with them; for example, there are potential concerns of cancer causing compounds created when grilling meats and may types of fish contain unacceptable levels of mercury.

A final issue is cost. Unfortunately, it's impossible to really discuss this aspect of protein choice because of variance in prices in different locations; what is inexpensive in one location may be very expensive elsewhere. As well, food prices change constantly due to fluctuations in availability; any price list provided in this book would be useless. In Appendix 2, I'll explain how readers can determine the relative cost per gram of protein for different foods and perform a cost comparison for foods available in their area.

Fowl

The category of fowl includes such foods as chicken, turkey, duck, goose and others. Like all animal proteins, fowl is a high quality protein. The fat content for proteins in this category can vary significantly. Cuts of meat such as the thigh can contain considerable amounts of fat while the breast meat of many types of chicken and turkey are essentially fat free if the skin is removed.

Additionally, many methods of preparation such as batter frying increase the fat content significantly; the generation of potentially carcinogenic compounds (discussed in more detail under red meat below) with certain modes of cooking is another possible concern. As mentioned above, the fat content of different parts of the bird can vary fairly significantly.

Fowl is a good source of the essential micronutrients zinc and iron. As discussed below, there appears to be a compound in both chicken and red meat that enhances the absorption of non-heme iron which can facilitate optimal iron intake.

From a practical standpoint, fowl can be eaten any time of the day although immediately before or during workouts would be inappropriate. Many athletes have trouble maintaining a proper diet when traveling or eating out; obtaining sufficient protein with a low fat content (if desired) can be difficult.

However, most restaurants (at least in the US) will have some type of grilled chicken breast meal or sandwich which makes getting sufficient lean protein fairly easy. Even the majority of fast food restaurants will have some type of grilled chicken sandwich. Coupled with a side salad or plain baked potato, this can allow athletes to maintain their desired training diet under these circumstances.

Fish

Fish is a high quality protein and many types of fish are extremely low fat, making it an excellent protein source choice. There are numerous varieties of fish and, similar to fowl, fish is appropriate at all times of the day except immediately around training. Of interest to dieting athletes, some research has found that fish keeps people fuller compared to either chicken or red meat (1).

Although fish is often chosen for its low-fat content, even the higher fat (typically cold water) fish typically contain most of their fat in the form of the omega-3 fish (w-3) oils.

These fats have a tremendous number of benefits, both for general health and athletic performance.

The beneficial effects of fish oils are numerous and include increased fat burning, decreased fat storage, decreased inflammation, decreased depression, mood stabilization, and decreased risk of heart disease and blood clots. More benefits are being found constantly and it's not an exaggeration to say that, if individuals were to take any single supplement on a daily basis, it should be a fish oil supplement. Although food scientists are working to increase the w-3 content of other foods (such as w-3 eggs described below), fatty fish remain the primary source in the food supply.

However, counteracting the potential benefits of fish and the w-3 fatty acids, there is increasing concern over the mercury content of many types of fish; because of their place in the food chain, mercury accumulates in the tissue of the fish to varying degrees. Chronic mercury exposure could potentially offset the benefits that fish might provide (1).

Because of the potential health effects of mercury, both the United States Environmental Protection Agency (EPA) and the World Health Organization (WHO) have set daily and weekly limits for safe mercury intake. Pregnant women, or women intending to become pregnant should minimize mercury intake from fish to avoid the possibility of birth defects.

A list of fish with their mercury concentration (in parts per million for a 3 oz serving) appears in Table 1 below.

Table 1: Mercury content (parts per million) per 3 oz serving

Fish	Mercury	Fish	Mercury
Whiting	ND	Shad	0.07
Ocean perch	ND	Whitefish	0.07
Scallops	ND	Pacific mackerel	0.09
Clams	ND	Cod	0.11
Shrimp	ND	Canned tuna (light)	0.12
Oysters	ND	Perch	0.14
Salmon (fresh)	0.01	Mackerel (Spanish)	0.18
Tilapia	0.01	Monkfish	0.18
Sardine	0.02	Snapper	0.19
Haddock	0.03	Sable	0.22
Crawfish	0.03	Halibut	0.26
Trout	0.03	Saltwater bass	0.27
Herring	0.04	Bluefish	0.31
Flounder/sole	0.05	Tuna (canned albacore)	0.35
Mackerel	0.05	Tuna (fresh)	0.38
Crab	0.06	Marlin	0.49
Pollack	0.06	Orange roughy	0.54

ND = non detectable levels of mercury

Source: Levenson CW and Axelrad D. Too much of a good thing? Update on fish consumption and mercury exposure. Nutrition Rev (2006) 64: 139-145.

Lower numbers are better, indicating less mercury per 3 oz serving of fish. On a daily basis, males should stay at the 0.19 level or less; fish higher than this level can be eaten up to twice per week. Females should stay at a level of 0.14 or less for daily consumption and can go up to 0.38 twice per week. Orange roughy, a staple of some dieting bodybuilders, is far in excess of acceptable mercury levels for regular consumption.

Red meat

Like all animal proteins, red meat is a high quality protein. Sources include beef, lamb, emu, bison, buffalo and many others. As discussed in Chapter 9, there are a number of negative health claims that are often made concerning red meat; for the most part these are more related to the high fat content that often accompanies red meat. Quite in fact, there are a number of good reasons for athletes to consume lean red meat on a regular basis.

Red meat contains high amounts of other nutrients including zinc, iron, B12, creatine and conjugated linoleic acid. Both zinc and iron are important to optimal athletic performance and many athletes are deficient in them (3,4). This is especially true for female athletes who are at much greater risk for iron deficiency and anemia due to blood loss during their menstrual cycle (4).

The iron in red meat is the most bioavailable form (called heme iron); heme iron is absorbed roughly ten times more effectively than the type of iron found in vegetable foods (called non-heme iron). As mentioned above in the section on fowl, a recent study suggests that some factor found in both beef and chicken improves the absorption of non-heme iron (5).

Creatine has a tremendous amount of positive research on it in terms of athletic performance and is discussed in detail in Chapter 12. The impact of conjugated linoleic acid (CLA) on human health is still a matter of some debate and is discussed in a bit more detail in the section on dairy foods below.

Simply put, in addition to its protein content, red meat is a nutritional powerhouse containing many nutrients of importance to both overall health and athletic performance. Given that most of the potential health problems associated with red meat can be avoided by consuming leaner cuts, I simply see no reason not to consume it regularly.

Humans are thought to have evolved on lean red meat and its consumption may confer numerous health benefits (6,7) in addition to providing nutrients crucial for optimal performance. By trimming the visible fat from red meat and choosing leaner cuts, red meat can and should be part of an optimal athletic diet.

The fat content of red meat can vary drastically from extremely low to extremely. For example, a lean eye of round contains roughly 4 grams of fat per 3 oz while short loin contains 20 grams of fat for the same 3 oz of meat; ground beef can vary as little as 4-8 grams of fat for very lean beef to upwards of 20-30 grams for the same 3 oz serving. Most game meat, such as buffalo and venison, is extremely low in fat, on par with skinless chicken breast.

Unfortunately, variance in local naming of meats, not to mention the huge number of possible cuts makes a list comparing the fat content of different meats nearly impossible. Additionally, fresh meats rarely, if ever, have nutritional labels which can make determination of the actual fat content more difficult. Readers can go to the USDA database to obtain information about specific cuts of red meat.

http://www.nal.usda.gov/fnic/foodcomp/search/

It's important to note a high fat content doesn't automatically make red meat bad, especially from the standpoint of heart disease. Not only total fat content but also the *type* of fat present in a food determines whether it will have positive, negative or neutral health effects.

Most foods, red meat included contain a mixture of different types of fats; quite in fact, while a small amount of the fat in red meat is thought to contribute towards heart disease, the majority present is actually neutral monounsatured fat with the remainder of the fats actually acting in a preventative fashion against heart disease (8).

However, given the potential concerns over large amounts of fat in the diet and other diseases such as colon cancer along with a general tendency to try to limit fat intake among many athletes, choosing leaner cuts of red meat is still probably the most prudent approach.

I should also note that some research suggests that adequate saturated fat intake is necessary to optimize testosterone levels (9); trying to eliminate all saturated fat from an athletic diet is probably a mistake in the first place.

Another concern regarding red meat (as well as chicken and pork) revolves around the production of various chemicals during cooking that may contribute to cancer risk. This is especially the case when meat is fried, broiled or barbecued and occurs to a greater degree when red meat is cooked medium-well or well done. Red meat cooked rare or medium-rare appears to cause fewer problems. Additionally, fewer of these chemicals are formed when meat is oven roasted or baked and almost none are produced when meat is stewed, boiled or poached. Athletes who consume large amounts of red meat should try to choose cooking methods that minimize the production of potentially dangerous chemicals.

I should probably also mention beef jerky which is a processed dried beef product that is typically extremely low in fat. Traveling athletes, who often have difficulty obtaining sufficient amounts of lean protein might consider jerky as a viable alternative. Containing plenty of lean protein, jerky doesn't require refrigeration and is easily transportable and can make obtaining proper nutrition while traveling much easier. One potential drawback to this is that jerky also tends to contain a large amount of sodium. Athletes should also be aware that jerky is not a pure protein source and typically contains a moderate amount of carbohydrates as well.

Pork

Pork protein includes products such as bacon, ham, pork tenderloin and others. Most of the comments I made above relative to fowl above more or less apply here: pork is a good high quality animal protein which can be consumed at any time except around workouts. The biggest drawback to pork is that many cuts are very high in fat. One notable exception is pork loin (or pork tenderloin) which is generally very lean.

The high sodium content of many pork products (especially processed foods) can also be problematic for athletes needing to make weight (or dry out immediately prior to a bodybuilding contest). Additionally, processed pork products such as spam, hot dogs, etc. tend to be full of any number of chemicals, in addition to being high in fat and sodium. Athletes should generally avoid them unless there is truly no other option available. Finally, there is a risk of trichinosis (a parasitic disease) with the consumption of undercooked pork.

Egg

Eggs have a long history in athletic nutrition, especially egg whites which have traditionally been consumed in vast quantities by bodybuilders. However, concerns about dietary cholesterol and saturated fat and their effect on heart health have scared many people away from whole eggs (the yolk contains all of the fat and cholesterol present in the egg).

It's important to note that the response to high intakes of dietary cholesterol in terms of changes in blood cholesterol is highly variable with most people showing no effect. Additionally, the changes which occur in blood lipid levels tend to be somewhat beneficial rather than negative, and whole eggs may contain other nutrients important for human health (10). Whole eggs appear to have received an undeservedly bad reputation nutritionally and even the American Heart Association has removed its recommendation to limit egg intake (11).

Like most animal source proteins, whole egg is very high quality; quite in fact, whole egg was once held as the "gold standard" for protein quality in humans. Additionally, eggs are typically inexpensive especially purchased in bulk. The various liquid egg products (such as Egg Beaters) are not quite as economical but may offer convenience over cracking numerous eggs.

At the same time, athletes wishing to reduce or limit total dietary fat intake often prefer to use egg whites (or liquid egg white products) as a concentrated source of nearly pure protein. While this reduces fat intake, and while egg whites are extremely versatile for cooking, egg white itself is not a tremendously high quality protein. As well, one individual egg white only contains about 3.5 grams of protein (and 15 calories) so it takes a lot of them to get an appreciable amount of protein.

A possible solution is to combine a number of egg whites with one or two whole eggs. This cuts fat intake while maintaining taste, and the higher quality protein in the yolk should ensure that the entire meal is of a higher quality. Alternately, consuming other high quality proteins in the same meal would help to prevent issues with the relatively poorer quality of pure egg whites. Athletes could add low- or non-fat cheese or some type of meat (i.e. chicken) to the egg whites and make an omelet; consumption of milk or yogurt with the egg whites should also improve the overall quality of the meal in addition to increasing the total protein content of the meal.

The consumption of raw eggs is not recommended due to a slight risk of salmonella poisoning. As well, uncooked egg whites are poorly digested and absorbed (12). Many companies now make a pasteurized liquid egg white that can be consumed raw or used for cooking although this still has the disadvantages of egg whites in terms of quality; the cost is typically higher than buying fresh eggs but such products offer greater convenience over cracking and separating egg whites.

Recently, eggs produced with a higher w-3 content have become commercially available. As sufficient intake of w-3 should be a part of every athlete's diet, individuals who dislike consuming pills or liquid fish oil may want to consider high omega eggs (or fattier fish, described above) as a potential source. While the cost is generally higher compared to regular eggs, this is offset by not having to purchase a separate w-3 supplement; I'd note that the content of w-3 in these eggs is not staggeringly high and it may be more cost effective to consume regular eggs and purchase fish oil in capsule or liquid form for supplementation. It's important to keep in mind that all of the w-3 in such eggs will be in the yolk; purchasing high w-3 eggs and only consuming the whites would defeat the purpose entirely.

Dairy

Dairy foods include such items as milk, yogurt, cheese, cottage cheese, ice cream and kefir (a cultured milk product). For semi-completeness, I would add condensed milk and even dried milk powder (which I'll discuss more in the chapter on protein powders).

All dairy foods derive from milk, typically from a cow (products made from goat and sheep are also available). Most dairy foods will have at least similar nutritional breakdowns because of this. I should mention that rice, soy, and almond "milk" are often used as a milk substitute but they share none of the qualities of dairy that I'm going to discuss below.

Nutritionally, dairy contains a number of compounds of interest and importance to athletes. In keeping with the theme of this book, clearly the protein content is of major importance. As I've mentioned previously, cows milk contains roughly 20% whey and 80% casein, two of the highest quality proteins both of which contain large amounts of the BCAAs and leucine (13). I've mentioned in previous chapters that high leucine intakes while dieting appear to stabilize blood glucose and improve the ratio of fat to muscle lost (14).

As discussed in Chapter 2 (and I'll discuss whey and casein protein powders in the next chapter), whey and casein have different functional characteristics in terms of their digestion speed with whey digesting rapidly while casein digests slowly. I'd note again that differences in digestion speed tend to be reduced when those proteins are consumed as part of a whole meal.

It's interesting to note that research suggests that the combination of whey and casein may be superior to either individually and, empirically, old school bodybuilders often advocated drinking large amounts of milk (usually whole milk) to gain muscle mass. Of course, athletes concerned with overall caloric or fat intake would tend to use lower fat or skimmed dairy products which are commonly available now.

However, the protein content of dairy is far from the only reason why it may be beneficial; dairy turns out to also contain a number of bioactive compounds of possible benefit (13). These include compounds native to both the casein and whey sub-fractions.

Research has found that these compounds can positively affect immune function, may lower blood pressure, and may have anti-microbial or anti-cancer properties (13,15,16). I want to note that most of these effects have only been shown in animal or test-tube models so far; whether or not they will apply to humans is still an area of debate and research.

With the exception of cheese, most dairy foods contain carbohydrate, primarily as the sugar lactose (milk sugar). Lactose may have its own biological effects, increasing calcium absorption for one (13). In addition, the fat content of dairy may have biological effects: for example, conjugated linoleic acid (CLA) has been examined for its role in cancer prevention and fat loss. While CLA generates impressive fat loss effects in lab animals, these effects have generally been minor or nonexistent in humans (17). Athletes choosing nonfat dairy will not be getting much CLA anyhow.

Fermented dairy products such as yogurt, kefir and others have been of some interest due to the presence of probiotics: live cultures which improve the health and function of the bacteria present in the human gut (13). Proper functioning of the gut and its bacteria can be affected by stress, antibiotic use and illness and inclusion of probiotics in the diet helps to correct this.

Arguably the primary role of dairy is as a calcium source and dairy products are the major source of dietary calcium in the food supply. A typical 8 oz serving of dairy contains roughly 300 mg of calcium (18). While other nondairy foods such as broccoli, beans and some cereals also contain calcium, it is generally not in the quantity found in dairy food. Additionally compounds such as phytates in grains and oxaloacetate in vegetables impair calcium absorption from those foods (18). Nutritionally, dairy foods are a calcium delivery vehicle optimized to provide calcium to the body.

Why is calcium important? Clearly one of the main reasons is bone health as adequate calcium intake is crucial for the development and maintenance of bone density (18). As discussed in Chapter 9, this is likely to be even more important in the face of high protein intakes where inadequate calcium intake can lead to bone loss. Additionally, calcium may play important roles in lowering blood pressure, reducing the risk of stroke, preventing colon cancer and reducing the risk of developing kidney stones (18).

Of more relevance to athletes, accumulating research indicates that dairy foods are important to regulate body weight and fat mass, partitioning calories from body fat to skeletal muscle (19). Early research suggested that this effect was due to the calcium itself and several different mechanisms of action were suggested. Research found that calcium itself might directly impact on fat cell metabolism by stimulating lipolysis, along with increasing fat burning in skeletal muscle. Finally, high calcium intakes appear to bind to dietary fat, carrying it out of the body without absorption.

However, later studies compared increased consumption of dairy foods to isolated calcium supplements and further additional effects on fat mass were often found; this suggests that some other aspect of dairy foods themselves might be having an effect.

One possibility is simply that dairy calcium is absorbed more effectively. There is also some indication of an interaction between the calcium and BCAA content of the whey portion of the protein. Finally, the bioactive compounds mentioned previously might be having an additional effect.

In any event, the research is clear that increasing calcium from dairy proteins (and this includes both whey and casein) can cause fat loss under both maintenance and reduced calories (19).

Which brings me in a roundabout way to some of the potential negatives of dairy products, some of which are real and some of which are most likely imagined. Oddly, given the research on dairy and fat loss, dairy products have had a reputation for "making people smooth" in the bodybuilding subculture. Contest dieters often talk of removing dairy from their diet weeks or months out from competition for this reason. The origin of this belief is a little bit obscure.

I suspect that some of it has to do with the fact that old-school bodybuilders often bulked up on copious amounts of whole milk; upon removing it from their diet (reducing calories and fat intake), they leaned out. Hence milk got a reputation as a food that kept people smooth (i.e. fat).

Another possibility, more likely in my mind, is that individuals on extremely low-sodium diets react to the sodium content of dairy foods by holding water. Cottage cheese for example, can contain up to 500 mg of sodium per serving with other dairy foods having 150-300 mg of sodium per serving. This could cause water retention for individuals who are restricting their sodium intake.

Another possibility is a true food allergy to milk as this could cause bloating. However, research shows that the true prevalence of allergies to cow's milk is only 1-3%, although it is often self-reported at 10 times that level (20,21). Clearly individuals with a true food allergy should avoid that food but it seems unlikely that all bodybuilders are in the 1-3% of clinically established cases. I suspect the sodium issue is the primary culprit here.

Of far more realistic concern with dairy is lactose intolerance (lactose mal-digestion) which is distinct from a true allergy (this topic is discussed in more detail at the end of the chapter). Lactose intolerance occurs when the body makes insufficient amounts of the enzyme lactase (which digests lactose); this causes stomach upset, diarrhea and gas in susceptible individuals.

Lactose intolerance often develops in adulthood although the prevalence varies by ethnic group. Generally speaking, darker skinned ethnicities (who typically stop drinking milk once they are past weaning) have a greater propensity towards lactose intolerance than lighter skinned individuals. I'd mention that lactose intolerance, like true food allergies, is often self-reported to a much greater degree than it actually occurs.

Athletes who suffer from lactose intolerance but who wish to consume dairy foods have several options available to them. Individuals with severe lactose intolerance often find that hard cheeses and yogurt can be consumed without problem; consuming dairy with meals can eliminate the problem as well (22). Regular consumption of fermented dairy products such as yogurt or kefir also appears to improve lactose tolerance, in addition to keeping the bacteria in the gut healthy (23).

Finally, lactose reduced or removed products are now available, companies also make supplemental lactase (in pill or drop form) that can be used to make digestion of dairy foods easier for individuals prone to lactose intolerance. Recently, a product containing certain strains of gut bacteria has been introduced to help improve lactose digestion, taken once per day in the morning, it claims to improve the ability to digest lactose.

Another possible downside of dairy is the fat content, which can be problematic for individuals trying to limit caloric or fat intake; the fairly high availability of low- and nonfat dairy products makes this a relative non-issue.

Between the extremely high quality protein found in dairy along with potentially beneficial bioactive compounds, added to the effect of dairy/calcium on body fat and body weight, there are tremendous benefits to be had from the regular inclusion of dairy foods in an athletic diet. This is true from both a health and practical standpoint.

Yogurt and cottage cheese an both be used as a carrier for protein powder (whey, casein or MPI can be mixed in to make a pudding type of dessert), a cup of milk can easily be added to other meals to bump up the protein content and quality, low- or non-fat cheese can be added to eggs or on top of lean hamburger meat, etc.

Specific to training, recent work has shown that the consumption of milk after resistance training promotes protein synthesis; oddly, whole milk worked slightly better than skim milk (24). Additionally, a recent paper found that skim milk generated a superior anabolic response compared to a soy based drink containing the same amount of protein (25). As discussed in Chapter 8, it appears that slower digesting proteins (or a mixture of fast and slow proteins) post-workout may generate a superior anabolic response.

Similarly, chocolate milk was recently shown to improve recovery following endurance training as well (26). Low-fat milk was also found to be superior to either water or Powerade for rehydration following endurance training (27).

About the only time that whole dairy foods would be inappropriate would be immediately prior to or during training as the slow digestion of casein could cause stomach upset. However, whey protein (discussed next chapter) can be consumed at those times.

While dieting for fat loss, it would appear that consuming dairy improves results in terms of greater fat loss and sparing of lean body mass loss. Athletes worried about potential problems with bloating or water retention may need to drop dairy out of their diets a few days or a week prior to their contest. Cutting dairy food 12-16 weeks out seems unnecessarily extreme, especially given that it may actually *hinder* optimal fat loss.

Legumes: Beans and nuts

The broad food category of legumes refers to both beans and nuts, of which there are many, many types. Soybeans (discussed separately below), kidney beans, pinto beans, lentils, chickpeas/garbanzo beans, etc. along with the myriad varieties of nuts all belong in this category. Many vegetarian athletes rely heavily on beans as a protein source and meat-eating athletes also often consume them as a supplement to their meat based protein.

In terms of their protein content, beans and nuts are considered a decent quality protein (depending on the method of determination) although their overall digestibility tends to be somewhat poor compared to animal based proteins (28). As mentioned previously, soy protein has been shown to be sufficient for meeting human maintenance protein requirements.

In addition to their protein content, beans tend to be high in fiber and other nutrients including zinc and calcium. Most beans are generally low in fat although nuts (with the exception of chestnuts), soybeans and chickpeas tend to contain more dietary fat. As well, nuts tend to contain the healthier fats and research has found that regular nut consumption may decrease the risk of cardiovascular disease and insulin resistance (29).

Interestingly, despite their high fat and caloric content, studies suggest that regular consumption of nuts either has a minimal effect on body weight or causes weight loss (30). This may be due to increased fullness from the nut consumption, a thermogenic effect from either the protein or fat content, or the fact that some of the calories in nuts go unabsorbed (31).

As far as negatives, I've already mentioned the relatively poorer digestibility of bean and nut protein relative to animal based proteins (discussed in more detail in Chapter 2). As well, the presence of anti-nutrients such as phytates, oxaloacetate and trypsin could impair the absorption of important nutrients such as calcium and zinc although this may be overstated (28).

Perhaps the biggest practical issue related to beans is that the fermentation of sugars in them can cause gas. This issue can be dealt with in a number of ways. One is to use a commercial produce such Beano which is available in both tablet and liquid. The tablets are consumed immediately prior to the meal while the liquid is added to gas producing foods before consumption to help avoid problems with maldigestion of the sugars.

Additionally, raw beans can be soaked overnight in water and baking soda; this serves to reduce the content of those same gas-producing sugars. For every 1/2 cup of beans, 2 cups of water and 1/2 teaspoon of baking soda should be added and the entire mixture left to soak overnight. The water is then drained off and the beans cooked and eaten with most of the offending sugars removed. Lentils and split peas seem to cause less gas problems for some people than other beans. Soaking does not appear to significantly impact on the nutritional content of beans.

Arguably the biggest problem with the higher fat beans and nuts is the high calorie and fat content. I'd note again that the fats present in these foods are typically the healthier fats; as well the relationship between nut intake and body weight is not cut and dry. Nuts, despite their high caloric and fat content, may help with weight or fat loss. That said, even small volumes of nuts can add tremendous numbers of calories and fat to the diet if they are eaten indiscriminately.

A high fiber intake is important to any athletic diet (except around workouts) for a number of reasons. Beans can be eaten with meals containing animal source proteins to provide extra protein, fiber and nutrients to the diet without adding significant amounts of dietary fat.

Soy

Soy products refer to foods such as soybeans, tofu, soymilk or soy protein isolate (discussed next chapter). Another soy food worth mention is texturized vegetable protein (TVP), a "meat substitute" made from defatted soy flour. High in protein and low in fat and sodium TVP can be used to increase the protein content of other foods or used in cooking as a vegetarian substitute for the meat that would normally be present.

The grand majority of comments I made above regarding legumes also apply to soy foods; however there is an addition issue related specifically to soy products that requires a separate discussion.

From a protein standpoint, soy protein is actually a decent quality protein; recall from Chapter 5 that it ranks equally to animal based products based on the PDCAAS scoring method. Studies have shown that soy protein is more than capable of supporting basic human protein requirements. However, like most beans, soy's digestibility is not fantastic (28).

Soy contains a reasonably large amount of the amino acids glutamine, lysine and the BCAAs; quite in fact some soy protein isolates contain more of these aminos than whey or casein. And while the fat content of soybeans is high (relative to some other foods), the fat content is primarily healthy fat.

The primary concern regarding soy products has to do with the phytoestrogen content and a great deal of controversy exists over this topic. Phytoestrogens are naturally occurring compounds in soy products that have weak estrogenic effects in the body. That is, they bind to the estrogen receptor and exert a weak signal (weaker than the body's natural estrogens would exert).

I want to note upfront that the research on the effects of phytoestrogens is highly mixed with both positive and negative studies existing; furthermore, whether or not phytoestrogens are a "problem" may depend on whether they are being consumed by men or women (32).

Most of the research into the effect of phytoestrogen has been performed on women. It's thought phytoestrogens might exert a beneficial effect in women by decreasing overall estrogenic stimulation. This occurs as the phytoestrogens binds to the estrogen receptor, sending a weaker signal than the bodies own estrogens would send. This could conceivably lower the risk of breast cancer. In peri- and post-menopausal women, phytoestrogens could have many potential health benefits related to bone or cardiovascular health (33, 34). Even there, scientific debate continues heavily as to the relative benefits or drawbacks of soy foods on human health.

Much less clear is the potential impact of the phytoestrogens on males, with concerns about the impact of phytoestrogens on male hormones (especially testosterone) being the primary concern. Interestingly, both negative and positive effects on hormone levels have been suggested (e.g. it's been theorized that raising estrogen in males might cause testosterone levels to go up).

Most of that research has examined the impact of phytoestrogens on prostate cancer; the idea being that by lowering testosterone levels in men, the progression of prostate cancer could be slowed. However, direct research into the topic has shown, at most, a small impact of soy products on the hormone levels of men (35).

One study, which gave men 70 mg/day of phytoestrogens, noted a slight increase in SHBG (sex-hormone binding globulin, a hormone that binds to testosterone) and slight decreases in free testosterone levels. Other studies using lower intakes (40 mg/day or less) have found no effects. Thus there may be a threshold level below which phytoestrogens have no measurable effect on male hormone levels. It's interesting to note that the average phytoestrogen intake in Asian communities (often claimed to consume a great deal of soy products) has been estimated at 15-50 mg/day.

The most recent research gave men 56 grams of soy protein per day for 4 weeks and examined hormone levels (36). The results were very mixed, one subject (who started with extremely high testosterone levels) had a massive drop in levels, several subjects showed a slight increase in testosterone.

Soy foods typically contain 1-3 mg of isoflavones per gram of protein so 56 grams of soy protein per day would have provided 56-168 mg of phytoestrogens per day, well above the threshold suggested above for hormonal effects.

At this point I consider the issue unresolved. Extremely large intakes of phytoestrogens daily may slightly impact on male hormone levels; smaller intakes appear to have no effect.

It is currently unknown what, if any, real world impact these small hormonal effects would have on male or female athletes in terms of performance, adaptation or recovery. Empirically, bodybuilders often used large quantities of soy protein (before other protein supplements became widely available) with no reported negative effects. Until more data is available, it seems reasonable to limit soy intake to one to two 20-25 gram servings per day (this amount includes soy protein powders, discussed in the next chapter). As mentioned previously, athletes should strive to consume a variety of proteins throughout the day in the first place in order to maximize benefits and minimize any potential negatives of any one source. With the large variety of proteins available for consumption, there should be no need for an athlete to overconsume any singular protein to excess in the first place.

I should mention that soy is being used as a protein source in many food items such as protein bars, "high-protein" cereals, low-carbohydrate foods and prepackaged meals; athletes who consume these foods as a regular part of their diet may already be ingesting some amount of phytoestrogens without realizing it. It's conceivable that adding soy foods (or protein powder) to such a diet could increase phytoestrogen intake to such a level to have a negative impact on hormone levels.

As a final comment, while early animal research suggested that soy protein might increase levels of thyroxine (T4, one of the thyroid hormones), the most recent research suggests that there is little impact in healthy individuals (37).

However, large levels of soy protein may impair thyroid medication uptake in individuals who are hypothyroid (37, 38). As well, it appears that the phytoestrogens in soy can cause hypothyroidism when iodine intake is insufficient (39). Given that iodized salt is one of the primary sources of iodine in the modern diet, and given that many athletes limit their salt intake excessively, excess soy protein could potentially cause problems.

Other proteins (grains, vegetables, fruit, etc.)

With a few exceptions such as many candies, which are pure sugar, and almost all oils, which are pure fat, most foods contain some protein. Vegetables and fruits contain trace amounts of protein at best, while many grains can contain quite a bit; for example, both a cup of oats and a large bagel may contain 8-10 grams of protein.

A question that comes up among obsessed athletes is whether these proteins should be counted in daily totals. It's usually pointless to include the gram or two that occurs in fruits and vegetables but the protein in grains can add significantly to total protein intake.

And while the protein in grains may not be the highest quality, assuming those grains are part of a mixed diet containing other high quality proteins, any issues regarding quality are both overstated and irrelevant. For this reason, I think that the protein in those foods can and should count towards an athlete's daily totals.

A final vegetarian protein that I want to mention is seitan, made from the gluten protein found in wheat flour. Seitan is high in protein with moderate amounts of carbohydrate and can be used for cooking or as a "meat substitute". Although I'll discuss food allergies in more detail in the next section, individuals with gluten intolerance issues should avoid seitan.

Food allergies and intolerances

To finish up this chapter, I want to make a few comments about food allergies and intolerances, which I touched on above in the section on dairy foods. Commonly, the terms food allergy and food intolerance are used interchangeably although they actually represent very different phenomena (40).

Food intolerance, such as lactose intolerance from dairy products, typically occurs due to a lack of appropriate digestive enzymes and this tends to cause upset stomach, gas, bloating or diarrhea. At worst, food intolerances typically cause some discomfort but no real danger.

In contrast, a true food allergy generates an immune reaction in the body. This is potentially much more severe and can cause respiratory, stomach, skin and cardiovascular symptoms; anaphylactic shock and death can also occur in extreme cases (40). True food allergies are typically caused when small amounts of proteins enter the bloodstream. This can occur during childhood when the gut lining isn't fully developed or later in life due to a compromised stomach barrier. Some allergens can also enter the body through the respiratory system.

While technically any food can cause a true allergic response, protein foods tend to be the most common culprits with milk, egg, peanuts, tree nuts, some fish and shellfish being the most common causes of allergies (41). Gluten, a protein found in grains such as wheat, barley and rye, is also a common source of food allergies (42). Gluten allergies can be especially troublesome for athletes with high caloric and carbohydrate requirements since grains cannot be consumed; increasing commercial availability of gluten free foods can help to ensure adequate calorie and carbohydrate intake.

True food allergies are thought to occur in 3-4% of adults. There are a number of different ways to determine the presence of a true food allergy but, from a practical standpoint, if eating a given protein source causes problems of the sort described above, that tells the athlete all they need to know. For the most part, there is little to no treatment for true food allergies; avoiding the problem food is the best and only option (40).

Summary

After reading this chapter, hopefully you realize that no single whole food protein can be considered the best. Whether it's protein quality, the presence or absence of other nutrients, or potential health risks, all proteins can be compared to one another in terms of their pros and cons.

For this reason, eating a variety of whole food proteins throughout the day would seem to be the best practical strategy; this should serve to minimize any negatives inherent to one source of protein while maximizing overall protein nutrition. Bodybuilders, especially when they are dieting, tend to consume the same single protein source (or alternate between two different proteins, usually chicken and tuna) and may be missing out on important nutrients.

Additionally, it seems possible that combining proteins in individual meals might confer some benefits. I've mentioned that consuming higher quality proteins can improve the quality of lower quality proteins and mixing and matching proteins at any given meal would seem to ensure that any deficiency present in one protein might be offset by something present in another protein. Vegetarians have done this for year by combining grains and beans but the same might very well hold for other proteins.

Various legumes (beans/nuts) or grains, providing both fiber and carbohydrate, could be consumed along with animal protein. Omelet's can contain ham, chicken or other meat along with low or nonfat cheese; a glass of milk or yogurt can always be consumed with other meals to provide the benefits of dairy in addition to the other protein present in the meat.

In Table 2 below I've provided a partial list of some common foods along with the protein content found in a given serving size.

Table 2: Protein content of some common whole foods

Food	Serving Size	Protein (grams)
Milk	1 cup	8
Cheese	1 oz	7
Yogurt	1 cup	8
Beef	1 oz	8
Chicken	1 oz	8
Fish	1 oz	7
Whole egg	1 large	6.5
Egg white	1	3.5
Beans	½ cup	7
Tofu	½ cup	10
Peanuts	¼ cup	9
Bread	1 slice	3
Broccoli	½ cup	2
Banana	1	1

As mentioned earlier in the book, animal source foods tend to be the most concentrated sources of protein. A typical serving of meat might range from 3-4 oz (about the size of a deck of cards) to double or triple that. This would provide anywhere from 21-28 to 60 or more grams of protein. In contrast, vegetable source foods contain far less protein for a given volume. Vegetables and fruit contain only trace amounts.

11

Protein Powders

Protein powders have long been an athletic staple, especially in the bodybuilding world, with amazing claims being made for them over the years. Even the general public has jumped on the protein drink bandwagon to some degree; this is especially true as emerging data has shown the benefits of higher protein intakes for both appetite control and weight loss.

While I feel that whole food should comprise most of an athlete's daily protein intake, protein powders can have advantages over whole food (1). For one, some bulk protein powders can be cheaper per gram of protein than many food proteins. Additionally, most protein powders are low- or nonfat and have excellent shelf life and stability. Because they can be exactly measured, protein powders can also provide easier nutrient control over the diet compared to whole food proteins.

Based on the rapidly emerging data discussed in Chapter 8, it seems fairly clear that protein powders are ideal around workouts, especially in the pre- and during-workout phases, where whole foods would cause stomach upset. Whether whole food proteins or powders are superior following workout is up to debate but protein powders can be beneficial there as well since many athletes find consuming liquid nutrients easier than eating solid food.

Outside of around workout nutrition, protein powders have a number of additional practical uses. Protein powders are often consumed as their own meal, either by themselves or with other nutrients (carbohydrates, fats, fiber). Meal replacement powders, containing protein, carbohydrates, fats, fiber and micronutrients are often used to provide easy and quick nutrition to support athletic performance.

A scoop or scoop and a half of protein powder can be mixed in various liquids (milk, juice, water, etc.) alone or with other nutrients to provide a quick and convenient 25-40 grams (or more) of protein. This can be beneficial for athletes trying to squeeze in sufficient protein around a busy work or school schedule. Athletes who have trouble meeting calorie (or protein) requirements from whole food often find that drinking at least some portion of their daily calories can make things easier; protein drinks can facilitate this.

As well, protein powder can be added to morning yogurt or cottage cheese to boost the protein content without adding extra carbohydrates (if any athlete is trying to avoid or limit them). Protein powder can even be worked into cooking recipes, added to pancake mix or muffins or other baked goods to produce high protein versions of those foods. Many athletes make up their own types of low-fat, high-protein desserts as well.

Simply put, between their convenience, cost and utility, protein powders are likely to remain a staple of diets for athletes forever; as discussed above, they certainly have many uses. However, I'd like to reiterate a point I made earlier in this book: because of the prevalence of their use, companies will continue to push expensive protein powders on athletes based on false promises and hype.

As an extreme example, a product has recently come to market that is essentially a purified hydrolyzed casein. With essentially no research supporting any benefits of this product, it retails for roughly 7 times as much as other protein powders.

Another recent trend in sports nutrition is to push protein hydrolysates based on the claim that the faster digestion speed will promote better anabolism or recovery following training. As discussed in Chapter 2, the difference in digestion speed between whey, casein and their respective hydrolysates are a few minutes at most; that's in addition to emergin research that faster digestion speeds may not be superior in the first place, especially after training. The presence of free form AAs in hydrolysates often makes them bitter tasting as well, on top of costing more.

Similar claims regarding the superior quality or absorption of one protein over another are often made, as discussed in some detail in Chapter 5. However, given the extremely high quality and digestion efficiency of proteins such as whey and egg to begin with, it seems extremely unlikely that other proteins will show significant differences in these regards.

Unless an athlete is dieting strictly and has to watch total food intake, eating a few percent more of a cheaper protein works just as well as consuming less of a much more expensive protein. The same holds true for hydrolysates versus isolates. An athlete overly concerned with the few minute difference in digestion could just as easily consume their protein 5 minutes earlier, rather than paying more for a bitter hydrolysate.

In any case, in looking at protein powders, as with whole food proteins in the last chapter, the protein content or quality is only one potential aspect when examining their value. Possible functional effects (such as impacts on immune system or overall health) along with potential effects on performance are equally important in determining what protein powder might be optimal for a given application.

At the end of each listing, you'll often find my comments regarding a given powder's taste, texture, mixability, etc. I want to make it clear that these simply reflect my personal opinion based on taste testing a number of products, and that individual variance in taste and texture preferences certainly exist. In an attempt to standardize my comparison, I obtained commercial samples of a number of products from an online retailer, all of them were chosen with an identical flavoring system. Thirty grams of protein powder was mixed in 8 oz. of water for 10 seconds in a shaker bottle and I based my comments on tasting a small amount of the drink. Many protein powders mix and taste differently in other fluids (i.e. milk, juice) and that may make otherwise poor tasting protein powders drinkable. I'd strongly suggest that any athlete trying a new type of protein powder obtain a small amount (if possible) to test for taste, etc. Purchasing ten pounds of a nearly undrinkable powder is simply a waste of the athlete's money.

Forms of protein powder

As I discussed in an earlier chapter, protein powders come in one of three primary forms: concentrates, protein isolates, and hydrolysates. Protein concentrates are typically about 80% protein with 5-6% carbohydrate; isolates are generally 90% protein with minimal carbohydrate. Hydrolysates are simply proteins that have been predigested (partially hydrolyzed) by throwing some digestive enzymes into the mix.

Concentrates are the least expensive form of protein with the price increasing for isolates; hydrolysates are the most expensive. As mentioned above, hydrolysates are often bitter due to the presence of free form AAs in the mixture.

The main advantage of isolates over concentrates is the slightly higher percentage of protein relative to the total; the lack of extra carbs may be important for athletes who are restricting or controlling carbohydrate intake. Empirically protein concentrates seem to cause bloating in some individuals relative to isolates as well.

As discussed above and in other chapters of this book, the supposed advantage of hydrolysates is speed of digestion although research doesn't really bear this idea out. Not only is the difference small or non-existent, there is some indication that faster digestion isn't necessarily always better in the first place.

As I discuss some of the different types of protein powders below, simply keep in mind that most are available as concentrates, isolates or hydrolysates. I won't be making individual comments (for the most part) on, for example, whey concentrate versus whey isolate versus whey hydrolysate.

As a final comment, it's not uncommon to see products containing a mixture of different protein types, based on the idea that this might generate a better overall response. Some companies also allow customers to customize their own protein blend.

For example, a combination of whey and casein might be produced to provide both a fast and slow digesting protein; a third protein might be added in an attempt to get a different digestion speed or AA profile. Outside of the small amount of work discussed in Chapter 8 examining types of protein following training, the topic is poorly researched.

Other products contain a mixture of different forms of the same protein. For example, one product contains a mixture of whey isolates, concentrates and some hydrolyzed whey protein.

As with the last chapter on whole food proteins, I want to look at the currently available protein powders in terms of potential advantages, disadvantages and applications. I'll make further comments as necessary. As with whole food proteins, while cost is clearly another issue of importance in choosing a protein powder, price differences between locales make this an impossible topic to discuss meaningfully. Readers should refer to Appendix 2 to see how they can set up their own price comparison chart for locally available proteins (whole food or powders).

One issue of importance to athletes using protein powders has to do with the actual amount of protein per serving. As you'll see below, the presence of other macronutrients (especially fat) can affect the actual amount of protein found in a single serving of protein powder; both whole egg and hemp protein contain considerably less protein per serving for this reason.

As an additional issue, while protein powders will always list a given standard serving size and generally come with a scoop for measurement, differences in the density of the protein powder itself can drastically affect whether the scoop actually contains the listed amount. For example, a standard 30 gram scoop may or may not actually contain 30 grams of protein powder depending on the density. Athletes may wish to obtain a digital food scale to actually measure out 30 grams of powder to see how it actually correlates with the included scoop.

Egg

Commercially, both egg white and whole egg powders are now available; the whole egg powders contain both the fat and cholesterol content of whole eggs themselves. I'd note that this significantly reduces the actual protein content per serving for whole egg protein. I don't have much to say about these types of proteins that I didn't cover in the last chapter on eggs themselves.

Egg white by itself is not a fantastic protein although whole eggs are extremely high quality. Egg protein powders typically don't mix well and have a reputation for giving users very smelly gas.

The primary use of egg protein powders would simply be convenience. Eggs are generally cheap and easy to obtain although cracking 6-12 of them at a time can be a hassle; an egg based protein powder used by itself (or mixed with other proteins) could provide an easy protein source in this situation. Additionally, athletes who cannot tolerate milk proteins could use egg protein powder as a substitute. Although I'm unaware of data on protein powder, egg appears to be a slowly digesting protein (as per Table 2 in Chapter 2) and would be most appropriate post-training.

Egg white powder mixes well although, mixed in water, it has somewhat of a thin texture. In contrast, whole egg powder mixes somewhat less well but provides a much thicker texture, most likely due to the presence of dietary fat.

Dairy: Whey, casein and milk protein isolate

I discussed dairy foods in the last chapter but isolated dairy proteins are also available in supplement form; they are currently of great interest in the sports nutrition world. As mentioned last chapter, whole dairy foods such as milk typically consist of 20% whey and 80% casein and most of what I discussed last chapter also applies to dairy based protein powders.

Both whey and casein are available as isolated powders and milk protein isolates (containing both whey and casein but without the tagalong carbohydrates and fats) are also available. Most commercial products have had the lactose content reduced or removed, making them appropriate for individuals with lactose intolerance. Athletes with true dairy allergies will still have to find an alternative.

The least expensive casein powders are typically in the form of caseinates, while a far more expensive form of casein called micellar casein is also available. Casein hydrolysate is available and, although it was used in one study described below, I question its utility; the primary benefit of casein is its slow digestion, hydrolyzing casein in an attempt to increase digestion speed seems to eliminate casein's primary advantage. I would note again that the actual difference in digestion time between casein and its hydrolysate is very small in the first place; any extra cost of hydrolyzed casein would appear to have no real benefit in the first place.

Whey comes in a number of different forms, defined by the amount of processing that has been done. Concentrates are processed further into isolates while cold filtration and ion-exchange separation attempts further purification without damaging the protein; price generally goes up significantly as you move from one type of whey to the next. Whey hydrolysates are also available; as discussed previously, there appears to be no difference in digestion speed and I see little benefit to be had.

Milk protein isolate (MPI) is simply isolated milk protein and contains 20% whey and 80% casein. To my knowledge MPI is only available as an isolate; concentrates and hydrolysates are not available. Recently, a complete milk dairy isolate (CMDI) has become commercially available. Similar nutritionally to MPI (although more expensive), CMDI is claimed to have been processed more gently.

As an additional option, skim milk powder is an extremely inexpensive way to add both high quality protein and calories to a diet. However, the lactose is still present (causing problems for individuals with lactose intolerance) and the carbohydrate could be a problem for individuals trying to limit their intake.

Due to their high digestibility and AA content, all dairy proteins are extremely high quality proteins as discussed back in Chapter 5. I'd mention again that, despite the occasional claim to the contrary, the biological value of whey does not and cannot exceed 100.

As discussed in previous chapters, whey and casein, despite coming from the same source are not identical. They have slightly different AA profiles with whey typically containing the highest BCAA content (23-25%) of any protein; casein comes in around 20% BCAA content. Casein also contains more glutamine than whey, soy or egg protein. I've mentioned research in earlier chapters suggesting that high intakes of BCAAs/leucine may offer benefits to both protein synthesis and while dieting (2,3).

As well, and as discussed in previous chapters, whey and casein show distinctively different patterns in terms of their digestion rate and metabolic effects on the body (4).

Recapping briefly, whey protein tends to digest quickly, flooding the bloodstream with AAs. While this tends to promote protein synthesis (with only a minimal effect on protein breakdown), it can also result in a greater oxidation (burning) of amino acids. As discussed in Chapter 7, this may be because excess AAs overload the body's ability to utilize them, pushing more towards oxidation. I'd note again that adding other nutrients such as carbohydrate, fat and fiber to whey makes it act much more like a slow protein; the other nutrients also decrease protein breakdown.

In contrast, casein forms a clot in the stomach, slowing digestion. While blood AAs don't increase to as great a level as with whey, casein ingestion generates a sustained low level of blood AAs for up to 7-8 hours. Because of this, casein tends to have a minimal impact on protein synthesis but decreases protein breakdown.

At first glance, this would suggest that whey is more of an anabolic protein (and hence better for mass gains) while casein would be superior while dieting (because of its anti-catabolic effects). What very limited direct research exists would appear to support this idea.

In one study, obese policemen were put on a diet and resistance training program supplemented with either whey protein or a casein hydrolysate; the casein group lost more fat and gained more lean body mass (5). In contrast, a recent study had lifters supplement their diet with either 1.5 g/kg of whey or casein while on a progressive resistance training program; the whey group gained significantly more lean body mass than the casein group (6). As discussed in previous chapters, a mixture of both proteins might yield superior results.

As mentioned above, the high BCAA/leucine content of both whey and casein are beneficial while dieting. Whey, casein and MPI are all good sources of dietary calcium which, as I discussed last chapter, seems to have a beneficial effect on body composition. From an appetite standpoint, at least one study suggests that whey may decrease appetite due to its effects on gut hormones (7); casein might have benefits in these regards due to its slow gastric emptying.

Although not yet studied, MPI might be superior in terms of appetite control by combining the hormonal effects of whey with the slow digestion rate of casein. Empirically, many people report that MPI keeps them full longer while dieting.

In terms of performance, subjects supplemented with either whey (as Immunocal) or casein made greater performance improvements with the whey; antioxidant levels were also increased to a greater degree (8). The latter effect is thought to be due to the high cysteine content of whey which may serve to raise tissue levels of a powerful antioxidant compound called glutathione.

In endurance athletes, an intake of 1.0 g/kg of whey protein per day has been shown to prevent the normal drop in glutathione status over 6 weeks of training (9). By maintaining tissue levels of cysteine, whey protein might also help to prevent muscle loss; this has been shown with N-acetyl-cysteine supplementation in individuals with low blood glutamine levels (10). From a more general standpoint, both whey and casein contain bioactive fractions that may confer health benefits to the immune system, gut and more as discussed last chapter (11,12).

As discussed in previous chapters, it's been suggested that a combination of whey and casein might give optimal results by combining the anabolic effect of the whey with the anti-catabolic effects of casein (13). In keeping with this, a recent study found that skim milk provides a superior anabolic response to soy protein (a fast protein), probably due to the more slowly digesting nature of the casein (14). Related to this, a recent paper found that whey alone was inferior at maintaining an optimal anabolic response compared to a total milk protein containing both whey and casein (15). This goes back to an idea I examined in Chapters 2 and 7, faster digestion speed may not automatically be superior under all conditions.

As I mentioned in the last chapter while discussing dairy foods, it's interesting to note that drinking copious amounts of milk was common among old school bodybuilders for muscle mass gains. MPI, containing 20% whey and 80% casein would provide this protein profile although bulk whey and casein could easily be mixed in different proportions. Alternately, whey could be added to milk, yogurt or cottage cheese to bump up the proportion of whey, with the dairy food providing casein.

Although the impact of MPI per se on muscle mass gains has not been examined, one study that found a mixture of whey and casein to be superior to whey, glutamine and BCAAs after training (16). Another found that milk (both skim and whole fat) stimulates protein synthesis following training (17).

As discussed in the chapter on nutrient timing, whey protein is more appropriate than casein or MPI before and during training; the casein fraction simply digests too slowly and may cause stomach upset. After workout, as mentioned, both whey and casein promote protein synthesis and a mixture of the two or MPI might be expected to promote superior results to either in isolation.

Whey is easily used in protein drinks due to its superior mixability and consuming fast digesting whey drinks in-between solid meals (as discussed in the chapter on meal timing) might provide some additional benefit in terms of protein synthesis for athletes trying to gain muscle mass.

From a practical standpoint, commercial whey proteins are generally highly soluble and mix easily. Whey can be added to almost any beverage (and some foods) and mixed with a spoon. A scoop of whey in a shaker bottle can be mixed rapidly and easily.

Casein, in contrast, tends to have somewhat of a chalky/bland taste and mixes poorly; a spoon is rarely sufficient with a shaker bottle or blender being required. Hydrolyzed casein mixes fairly well although it tends to froth up during the process; I would describe it as undrinkable due to the bitter taste. Even a small amount was gag-inducingly bad and I find it unlikely that most athletes would drink it on a regular basis.

Similar to casein, MPI which tends to mix poorly without a shaker bottle or blender; it also tends to have the same chalky taste as pure casein. The newly available complete milk dairy isolate (CMDI) has a less chalky taste and mixes more easily than MPI.

Soy isolate

A great deal of what I discussed in the last chapter for soy foods applies to soy protein isolates and I won't repeat that information here. However, soy protein isolates are typically fortified with limiting AAs to increase their overall quality compared to whole soy foods. Interestingly, compared to whey, casein, eggs and beef, soy protein isolates can actually be higher in what is termed the critical cluster of AAs: glutamine, BCAAs and lysine.

The low cost of soy protein isolate makes soy protein powders attractive. Vegetarian athletes, or athletes who can't consume dairy proteins because of allergy, might consider soy protein powder as an alternative.

In terms of exercise performance, two studies have shown that 40 grams of soy protein per day improved antioxidant status to a greater degree than whey when combined with resistance exercise (18, 19). And while one study showed no difference in lean mass gains for soy versus whey protein over 6 weeks (20), a longer study of 12 weeks found a greater trend for LBM gains in a group consuming milk protein rather than soy (21, 22). Soy may not be an ideal protein for promoting lean body mass gains from resistance training.

Related to this, a recent paper found a soy drink to be inferior to skim milk for promoting protein synthesis following resistance training (14). This is probably related to the fact that soy is a fast protein like whey. As well, research has shown that soy digests rapidly and may be used preferentially by tissues in the gut whereas dairy proteins provide more amino acids to skeletal muscle (23,24). Adding other nutrients to soy such as carbohydrate, fat and fiber might be expected to improve soy's performance in this regard by slowing it down; this has not been tested.

As I mentioned last chapter, it's currently unknown what, if any, impact the intake of soy protein/phytoestrogens actually has on male athletes in terms of muscle growth, adaptation or recovery. Empirically, soy proteins have been in use for several decades now and reports of problems simply aren't there; while this certainly doesn't prove that excessive phytoestrogen intakes couldn't cause problems, there seems little reason for the type of alarmism that so often surrounds the topic.

Once again, until more research is available, it still seems prudent to limit soy protein isolate intake somewhat. Last chapter I suggested that limiting total soy protein intake to 1-2 servings daily seemed prudent and I'd simply reiterate that comment here.

As I mentioned in the last chapter, many food manufacturers are now using soy protein to boost the protein content of many pre-packaged foods and athletes may already be consuming a good bit of soy protein without knowing it if these make up a major part of their diets. I encourage athletes to check the ingredient list on the foods they are eating to see if soy already comprises a large proportion of their protein intake before consuming additional amounts.

The fast digestion rate of soy might make it appropriate for pre- or during workout purposes, no research has examined this to my knowledge. Like whey, soy could conceivably be added to other foods to bolster the protein content; again, potential negatives from very high phytoestrogen intake are a concern here.

Soy protein isolate mixes well although it is thick and somewhat gritty. The taste is fairly neutral although it has a hint of "nuttiness".

Other vegetarian proteins: Pea, rice and hemp protein

A number of other protein isolates are available for vegetarian athletes who cannot consume egg or dairy proteins, these include pea, rice and hemp protein isolates.

Pea protein is, as the name suggests, isolated from peas; it provides a decent quality vegetarian protein source. It has recently become available and is fairly inexpensive and the taste is passable. Pea protein does not mix fantastically, even in a shaker bottle, and is very thick. The taste is distinctive although I'd have difficulty describing it.

Rice protein is isolated from white or brown rice and has an amino acid profile similar to soy protein at a similar cost. I was unable to taste test rice protein.

Finally is hemp protein which is isolated from hemp seeds. Hemp is a surprisingly good quality protein and hemp protein isolate contains both fiber and essential fatty acids. However, it is relatively expensive and the fiber/EFA content lowers the protein content per serving. Athletes wishing to increase both their fiber or EFA content might consider hemp protein although, cost-wise, it's probably less expensive overall to get a less expensive protein concentrate and simply supplement with fish oils. Hemp protein mixes fairly poorly leaving a sediment at the bottom of the bottle, the taste is somewhat grainy because of this; the taste is passable although, like pea protein, it has a distinct taste.

Summary

While I feel that athletes should rely on whole foods for the majority of their protein needs, it's clear that protein powders can offer a number of benefits in terms of price, convenience, dietary control, etc. Additionally, for around workout nutrition, especially immediately before and during training, protein powders are preferred over whole protein foods to avoid stomach upset.

Protein powders are generally available as concentrates, isolates and hydrolysates with price increasing as more processing is done. Isolates contain more protein per serving (and less carbohydrate and fat) than concentrates. Hydrolysates have been partially predigested; this can give them a bitter taste due to the presence of free form AAs. The purported benefit of hydrolysates is faster digestion but research does not generally bear out a significant difference between the digestion rate of proteins and their hydrolysates. Almost no research has examined whether protein hydrolysates generate a better anabolic response than the cheaper isolates and there is some indication that slower digestion speeds may actually be superior in the first place.

As with whole food proteins, it should be clear that no protein powder can be considered best in absolute terms; all have their individual pros and cons. While I personally see little benefit to egg protein powders, they may be beneficial as a high quality protein for athletes with intolerances to dairy.

While whey has been the current top protein (at least commercially), it should be clear that casein and MPI/CMDI offer significant advantages due their slower digestion time. This advantage may be offset by poorer mixability and a chalky taste. Slow proteins are also not optimal before or during training.

Soy protein isolate is still embroiled in a great deal of controversy; while the AA profile of soy isolate is good, concern over phytoestrogens and questions regarding their ultimate biological effects are still up to debate. My feeling at this point is that soy protein powder can be used (and may be one of few options for vegetarian bodybuilders) but its total intake should probably be limited to some degree. For athletes who can digest them (i.e. no dairy allergies), dairy proteins would still appear to be superior.

Rice protein isolate would seem to have little to recommend it, a combination of high cost and poor taste makes it a poor choice except for athletes who simply can't tolerate anything else. Pea protein isolate has recently become commercially available, a low cost and decent mixability and taste make it an appropriate vegetarian protein source. Hemp protein, while expensive, is a decent quality protein and contains both fiber and EFA's, its mixability is poor leaving a sediment in the bottom of the bottle, its taste is also somewhat grainy as well.

12

Supplements

Having discussed whole food proteins and protein powders, I want to talk about certain individual amino acids (or mixes of amino acids) or protein based supplements that are marketed towards athletes. I want to make the point again that anything described in this chapter should be considered only *after* overall protein intake (including both whole food proteins and powders) has been taken care of. For an athlete to worry about their glutamine or BCAA status when they are not getting sufficient dietary protein in the first place is going about things in the wrong way. Food comes first, supplemental protein powders second, any individual supplement should be considered last.

I bring this up for at least two major reasons. First and foremost, it's fairly ridiculous to worry about supplements that might add, at best, a small benefit when an athlete doesn't already have their basic nutrition and training program in line. Additionally, as you'll see in my discussion below, many of the studies which found a benefit of various amino acids were done with athletes consuming insufficient protein in the first place. When protein requirements aren't being met in the first place, it's far more likely that an individual nutrient might provide a benefit; however when protein requirements *are* met, this is far less likely to be the case.

I can't possibly discuss every product in existence; instead, I'm going to try and focus on the products that are currently popular with athletes and/or which have some research to support them in some way (or have research showing that they are ineffective). As with most supplements, a great many products have come and gone over the years and this is likely to hold true for many of the current crop of products.

In this chapter, I've tried to examine a variety of supplements from both a theoretical and applied standpoint. In referencing each product, I've focused as much as possible on research in otherwise healthy humans. A great deal of supplement research tends to be done using nutrient infusions, with animal models, in various disease states, or in vitro (meaning it is done in a test-tube or petri dish); unfortunately these models tend not to transfer well to healthy athletes consuming those nutrients orally.

While I am always optimistic that truly beneficial performance supplements will come along (and at least one, creatine, is discussed thoroughly below), the supplement industry has a rather poor track record in these regards. Supplements appear surrounded by immense amounts of hype, generally based on poor research models, only to disappear a few months later, then to be replaced by the newest magic pill of the month.

Glutamine

Glutamine is perhaps one of the most popular of the individual amino acids, although this reputation is not entirely deserved. While not an essential amino acid, glutamine is one of the most prevalent aminos in the body and can become conditionally essential (meaning that the body cannot produce enough and it must be obtained from the diet) under conditions of trauma, sepsis or burn. Please note that even the most intensive training doesn't even come close to damaging the body as much as these types of injury.

One of the oddities about glutamine metabolism is that glutamine is used to a great degree by the gut and most oral glutamine never reaches the bloodstream in the first place; as much as 65-75% of ingested glutamine will be used preferentially by the gut (1). In this regards, glutamine may have some benefit for gut healing in conditions such as irritable bowel syndrome (IBS). Glutamine obtained a reputation as a muscle-building compound when it was discovered that adding glutamine to isolated rat muscle increased protein synthesis and inhibited protein breakdown (2,3). It's important to note that the increase in skeletal muscle glutamine concentration in those studies was about 10 fold, a level of increase that simply doesn't occur in humans with supplementation (1). Studies using both intravenous infusion (4) and oral intake of glutamine (5) have found no impact on human skeletal muscle protein synthesis, with little to no change in skeletal muscle glutamine concentrations.

What little direct research exists on glutamine as an anabolic suggests that it is ineffective. One study gave subjects either 0.9 g of glutamine/kg lean body weight per day (81 grams of glutamine for a 100 kg athlete at 10% body fat) or the same amount of maltodextrin (a carbohydrate supplement) over 6 weeks of resistance training (6); no difference in strength or lean mass gains was seen between the groups. Another study gave subjects 0.3 g/kg of glutamine (30 grams for a 100kg athlete) prior to weight training and found no improvement in weightlifting performance (7).

In a very recent study, subjects were provided with either carbohydrate and essential amino acids or carbs/EAAs with added glutamine following endurance training; the addition of glutamine had no impact on post-training anabolism (8). Despite numerous claims to the contrary, glutamine appears to have no real benefit to muscle building, performance, or recovery following training.

Glutamine has also been suggested to be useful for fat loss, for a number of potential reasons. One study found that an oral intake of 2 grams of glutamine elevated growth hormone (GH) and blood bicarbonate levels; this is discussed further in the section on GH releasers. Another gave subjects 0.25 g glutamine/kg (25 grams for a 100kg athlete) with a standardized meal; the glutamine increased post-meal energy expenditure and increased fat oxidation by 42 calories (9) representing roughly 4 grams of fat burned. While this might conceivably benefit fat loss, it seems a rather cost-ineffective way to do so. A 100kg athlete would require 75 grams of glutamine per day to burn an additional 150 calories of fat per day. A dietary fat reduction of 15 grams/day would accomplish the same.

It's also been suggested that glutamine might be anti-catabolic on a diet, sparing muscle loss; the mechanisms were discussed back in Chapter 6. However, research does not support this idea either. In the only study I'm aware of, wrestlers subjected to a large caloric deficit were given 1.5 g/kg of protein/day and either 0.35 g/kg (35 grams of glutamine for a 100 kg person) or placebo for 12 days; both groups lost equivalent amounts of body mass, lean body mass and fat mass (10); the added glutamine had no effect

The immune system also uses glutamine as a fuel and protecting immune system function may be one of its primary benefits when the training load is very high, especially for endurance athletes (11,12). In one study, endurance athletes were given 5 grams of glutamine immediately after and again 2 hours after competition; there was a significant decrease in the reported incidence of sickness (13). Other studies have not supported this finding however (14,15).

Low plasma glutamine levels have been linked with overtraining (16), making supplementation of potential benefit. While glutamine levels often fall during prolonged endurance training, they are either unchanged or go up following short-term, high-intensity training (17). This suggests that endurance athletes might benefit more from glutamine supplementation than strength/power athletes. Athletes who perform both heavy weight work and endurance training (rugby, football, etc.) might also benefit from glutamine due to their high training volume.

As noted in Chapter 6, muscle tissue actually synthesizes glutamine from other amino acids. At least one study suggests that BCAA supplementation can protect immune system function in endurance athletes under heavy training (see BCAAs below) and glutamine intake may limit the body's need to break down BCAAs in the first place (18). Sufficient carbohydrate intake (30-60 g/hour) during endurance training also limits the decrease in muscle glutamine levels (19); the nutrient recommendations given in the nutrient timing chapter will help to protect the immune system.

In some studies, glutamine or glutamine peptides (from wheat protein) has also been used to refill muscle glycogen (20). This seems an expensive and inefficient way of doing things; carbohydrates are cheaper, taste better and work just as effectively. The nutrient recommendations given in Chapter 8 for post-workout nutrition should optimally refill muscle glycogen in the first place.

Application

Glutamine is relatively inexpensive and may be useful in doses of 5-10 grams/day to protect immune system function during periods of high-volume training. This is true for both endurance athletes and strength/power athletes who engage in some form of endurance conditioning during their training.

As discussed below, BCAAs may work more effectively in the first place and athletes consuming the amount of protein recommended in this book should be getting plenty of BCAAs to begin with. Empirically, many have found that taking high doses of glutamine (10-20 g/day) with vitamin C (multiple gram dosing) helps to fight off minor infections.

Much higher doses of glutamine or gluatamine peptides are required for glycogen compensation but I don't recommend this. Carbohydrates or a carbohydrate/protein mixture work just as effectively post-workout to resynthesize muscle glycogen. Glutamine is ineffective as an anabolic compound, even at massive doses. Nor does it appear to improve acute training performance or spare muscle loss on a diet.

Glutamine peptides, derived from wheat protein, have now become available; they are roughly twice the cost of straight glutamine in commercial products. Outside of the glycogen synthesis study discussed above, I am unaware of any research into glutamine peptides and athletic performance or muscle growth.

Tyrosine

Tyrosine is an amino acid that acts as a precursor substrate for adrenaline, noradrenaline and dopamine in the brain (21); it has been examined from a performance standpoint for this reason. By itself tyrosine doesn't appear to have any major impact on performance, even in high doses (22). Tyrosine also failed to improve endurance performance when combined with carbohydrate (23).

However, empirical evidence suggests that tyrosine can have an impact as a pre-workout (or pre-competition) stimulant when combined with other substances; tyrosine appears to act synergistically with the other compounds.

Athletes combining 1000-3000 mg (1-3 grams) of l-tyrosine with 200 mg of caffeine and a high glycemic index carbohydrate (the insulin helps to drive nervous system output) have noted stimulant effects similar to the ephedrine/caffeine (EC) stack; performance increases of 5% or more are not unheard of.

Athletes who may be drug tested or who will not (or can not) use the EC stack may wish to consider this combination before intense workouts or competition. While not tested in humans, animal studies have also found that tyrosine can potentiate the effects of the EC stack (24).

Application

Tyrosine may be useful as a non-ephedrine based pre-workout stimulant when used with caffeine and/or to potentiate the ephedrine/caffeine stack (for either dieting or performance purposes). One to three grams of l-tyrosine with 200 mg of caffeine and a small amount of sugar approximately 30-60 minutes prior to training or competition seems to be effective. By itself, tyrosine appears to be relatively useless.

Branched chain amino acids/leucine

The branched chain amino acids (BCAAs) are so-named because of their branching chemical structure and are comprised of the essential amino acids leucine, isoleucine and valine. As discussed in Chapter 2, the BCAAs are unusual in that they are not metabolized to any significant degree in the liver, most ingested BCAAs make it to the bloodstream to be used by skeletal muscle.

As discussed previously, BCAAs can be used for energy directly in the muscle and this effect is more pronounced when muscle glycogen has been depleted. The BCAAs are found in high quantities in all high quality dietary proteins, averaging 15-20% of the total protein content. Whey protein may contain 23-25% total BCAAs.

A typical diet containing high quality protein will provide 15-20 grams of BCAAs for every 100 grams of protein ingested (25); diets containing a significant amount of whey protein will contain slightly more than this. A 100 kg athlete consuming 3.0 g/kg protein, or 300 grams of protein per day, would be expected to be consuming 45-60 grams of BCAAs per day; again, this value would be slightly higher if a large amount of whey protein was being consumed.

I bring this up for reasons mentioned in the chapter introduction; the impact of supplemental amino acids when protein intakes are very high may be completely different compared to their use when dietary intake is insufficient to begin with.

The BCAAs have been extensively studied for a number of roles relevant to athletic performance. This includes sparing muscle mass while dieting, improving immune system function, limiting fatigue, promoting protein synthesis and others.

As I've mentioned in previous chapters, recent studies have found that diets high in protein and specifically leucine may help with blood glucose stability and protein sparing on a diet (26). As mentioned in an earlier chapter, both whey and casein are high in leucine; as discussed in Chapter 11, both may have potential benefits while dieting for this and other reasons.

Related to this, an early study in wrestlers found that the provision of extremely high quantities of BCAAs while dieting limited lean body mass loss and increased visceral fat loss (27). In that study, wrestlers were given 24.4 calories per kilogram (1660 calories for 68 kg/150 lb individual) with 20% protein. This would have only provided 80 grams of protein per day or roughly 1.2 g/kg (0.5 g/lb) of protein. The supplement group added an additional 52 grams of BCAAs to their diet; this spared LBM and generated a slightly greater fat loss.

But let's consider this within the context of the protein recommendations in this book. A strength/power athlete such as a wrestler has a recommended protein intake of 3.0-3.3 g/kg (1.4-1.5 g/lb). A 68 kg athlete at this protein intake would be consuming between 204-224 grams of protein per day, far more than used in the study. Assuming an average BCAA content of 20%, the additional 120-140 grams of protein would already be providing 24 to 28 grams of BCAAs from the protein alone; if whey made up some proportion of that total, even more BCAAs would already be present in the diet. Would supplemental BCAAs still have had a benefit or were the wrestlers simply receiving insufficient protein in the first place? I'm inclined to think it was the latter.

As mentioned above, BCAAs are heavily involved in the function of the immune system (28); one study found that the provision of 6 g/day of BCAAs for 15 days decreased the incidence of infection following a triathlon (29). As mentioned above, BCAAs and glutamine metabolism are intertwined and BCAA supplementation may be a more effective way of protecting the immune system, as this tends to protect the body's glutamine stores. I'd mention that endurance athletes often under-consume protein in the first place, a single 25 gram serving of whey protein would provide that same 6 grams of BCAAs found to improve immune system function.

Another possible role of BCAAs for endurance athletes is in preventing what scientists refer to as central fatigue (30). The earliest hypothesis for central fatigue suggested that, during exercise, as levels of BCAAs in the blood drop, more tryptophan is able to enter the brain; this would be expected to increase serotonin levels, making athletes drowsy and unmotivated. By maintaining blood BCAA levels, this could be prevented, limiting fatigue during exercise. Controlled studies using various nutritional interventions, including BCAA supplementation have failed to demonstrate a consistent effect however (31).

More recent research suggests that central fatigue is far more complex, involving interactions between a number of different neurotransmitters and compounds in the body (32). One of importance to this discussion is ammonia which can contribute to central nervous system fatigue (31). I bring this up as studies have shown that BCAA supplementation can *increase* ammonia levels during exercise (33), potentially contributing to fatigue. Interestingly, this doesn't occur with whole protein supplementation during exercise due to the presence of ammonia binding amino acids found in whole protein (34).

This may explain some of the studies on BCAAs showing no benefit on performance; while the BCAAs may have impacted on the level of serotonin produced in the brain, increased ammonia levels could have negated any benefit. Whey protein consumed during training should actually be superior to isolated BCAAs in this regards; due to the presence of AAs that bind ammonia, whey not only provides BCAAs during exercise but also limits the increase in ammonia levels that contribute to fatigue.

In addition to its potential effects on LBM maintenance while dieting and immune system function, it's also been suggested that hard training athletes might require more leucine; both aerobic and anaerobic exercise along with strength training have all been shown to decrease serum and plasma leucine levels by 11 to 30% and requirements may be higher because of this (35). Given the presence of BCAAs in all high quality proteins, it seems likely that simply increasing dietary protein to the levels recommended in this book will more than meet those increased needs.

An early study in endurance athletes suggested that BCAA intake might modify the hormonal response to training, preventing an exercise induced drop in growth hormone and testosterone levels (36); this might provide a more overall anabolic state following training. BCAA supplementation (12 g/day for 14 days) has also been found to reduce markers of muscle damage following 2 hours of bicycle exercise (37); as discussed in

Chapter 8, it's becoming clear that the consumption of even small amounts of protein during endurance exercise can limit muscle damage so these effects are not surprising.

A recent study in swimmers found that 12 grams of BCAAs/day for 15 days resulted in significantly less muscle breakdown (measured by levels of 3-methylhistidine) following training (38). Another recent study, in outrigger canoeists, found a performance improvement from a rather small supplement of leucine (roughly 3 grams for a 68 kg/150 pound individual) but with a protein intake of 0.85 g/kg (0.4 g/lb), their protein intake was fairly low to begin with (39). Increasing protein intake to the levels recommended in this book would have provided far more leucine than the small supplement provided.

Of relevance to strength athletes, BCAA supplements have long been known to stimulate protein synthesis (40). Recent research has demonstrated that leucine probably plays the major role (41) via stimulation of a biochemical sensor called mTOR. For example, a recent study provided 0.1 g/kg BCAAs (10 grams BCAAs for a 100 kg athlete) during and after resistance exercise (42); the researchers found that this increased activity of several of the signaling pathways involved in hypertrophy. I'd point out that the protein recommendations given in Chapter 8 would easily provide this amount of BCAAs (along with the full complement of both essential and non-essential amino acids) in the first place.

As mentioned Chapter 8, a recent study found that the addition of 6 grams of leucine to 14 grams of protein and carbohydrates had a small impact on overall net protein gain with the effect most likely due to an increase in insulin levels from the additional leucine (43). As discussed in the timing chapter, that study was a little odd in that it gave the various drinks for 6 hours following training, hardly applicable to most situations. As well, the amount of protein provided was below the amount needed to provide a maximal stimulation of protein synthesis. I find it unlikely that adding a gram or two of free leucine to a single post-workout drink containing sufficient amounts of protein and high glycemic index carbs would have any benefit.

In another study, 50 mg/kg of leucine (5 grams of leucine for a 100kg athlete) given to strength/power athletes on a fairly low protein intake (1.26 g/kg) prevented the decrease in blood leucine levels of 10 weeks of training but had no impact on hormones or performance (44). I'd mention again that the protein intake used in this study is less than half of what this book recommends, simply increasing total protein intake to appropriate amounts would have provided significant amount of BCAAs and leucine without the need for supplementation. For the most part studies have not generally found an impact of BCAAs supplements on performance (45,46).

However, one of the most commonly cited studies in favor of BCAA supplements is an Italian study (not translated in full to English) that compared either 0.2 g/kg of BCAAs before and after training to a non-caloric placebo. That is, it compared high dose BCAAs to nothing around training. So a 100kg athlete would be consuming 40 grams of BCAAs around training (47). The BCAA group showed greater improvements in weight and in squat and bench performance and this has often been used as a justification for the consumption of massive doses of BCAAs around training.

Given the data I presented on the importance of pre-, during- and post-workout nutrition in Chapter 8, these results aren't surprising: consuming *anything* around training is generally better than consuming nothing and the studies using various combinations of nutrients around training show improved results compared to nothing at all.

A strength/power athlete following the nutrient timing recommendations in that chapter should be consuming somewhere between 60 and 100 grams of protein around training, from whey and/or MPI. That would provide between 15-25 grams of BCAAs in addition to the full complement of other amino acids. Whether additional BCAAs would provide more benefit is currently unknown.

For completeness, I should mention HMB (beta-hydroxy-beta-methylbutyrate), a metabolite of leucine. Early research found that HMB limited muscle damage and led to greater gains in muscle mass (48). Later research determined that HMB only had an impact in untrained individuals (49); trained athletes derived no benefit from HMB (50,51). Empirically, few have reported benefits from HMB at the recommended daily dose of 3 g/day. Some coaches feel that higher doses (6 g/day) coupled with extremely intensive training might unmask an effect of HMB on muscle growth or damage but this has not been studied.

Interestingly, HMB might have some benefit to endurance athletes, specifically runners (who impose a good deal of eccentric stress to their muscles). In one study, runners supplemented with 3 grams of HMB per day for 6 weeks showed lower markers of muscle damage following a 20 km run (52). A recent study also found that 3 g/day of HMB improved the adaptation to endurance training in untrained college students (53). However, as with the data on strength athletes, it appears that HMB only improves aerobic adaptations in untrained individuals; trained endurance athletes receive no performance benefits from supplementation (54).

Application

Even in bulk, effective daily doses of BCAAs can get fairly expensive. As mentioned above, while early BCAA powder supplements tended to have a foul taste, newer products are reported to taste much better. Because of the doses typically used, pills are neither cost effective nor practical.

Overall, BCAAs do appear to be quite important to overall health and immunity, as well as muscle growth (directly stimulating protein synthesis); they may also be of benefit for endurance athletes to protect immune function and/or hormone levels. Their role in the prevention of central fatigue is more questionable; isolated BCAAs might actually contribute to fatigue by raising ammonia levels.

The unanswered question is whether additional BCAAs have a demonstrable benefit when protein intake is set adequately in the first place; the majority of studies examining BCAAs supplementation appear to have done so on a background of insufficient protein intakes (at least relative to the recommendations being made in this book).

Consider the study cited above which found an effect of leucine given to strength/power athletes consuming a mere 1.26 g/kg of protein (0.6 g/lb). That's roughly half of the lowest recommendation made in this book for total protein; would extra leucine have had an effect if the subjects were consuming more protein in the first place? I doubt it.

As stated above, every 100 grams of high-quality protein will provide 15-20 grams of BCAAs; a strength athlete consuming 180-200 grams of high quality protein per day is already obtaining 30-40 grams of BCAAs in their diet. Additionally, if the nutrient timing guidelines made in this book involving the intake of whey before and during training and casein, MPI or even milk afterwards, the athlete will be consuming large amounts of BCAAs, along with all of the other amino acids around training.

BCAA supplements might have benefits during periods of strict dieting, or if protein intake falls below optimal levels. Some bodybuilders have experimented with sipping BCAA solutions during the day while dieting in an attempt to stave off catabolism; I suspect that simply ensuring a constant low level of amino acids in the bloodstream (through the use of slowly digesting proteins such as casein or MPI and an appropriate meal frequency) would have as much if not more benefit.

Endurance athletes who, as a whole, tend to under-consume protein (and generally don't use supplemental protein powders even though they should) might also benefit from additional BCAAs to protect immune function. Frankly, I'd rather see endurance athletes ensuring sufficient protein intakes in the first place, rather than try to make up for a dietary inadequacy with a supplement. Consuming sufficient carbohydrates during training is also crucial for maintenance of proper immune system function.

As noted, HMB might have some benefit for endurance athletes, especially runners, at a dose of 3 grams/day to limit muscle damage; improvements in endurance performance effects seem to be limited to untrained individuals. For strength/power athletes, HMB has proven ineffective as a supplement, at least at the recommended dose of 3 g/day. Higher doses (6 g/day) might have some impact during periods of extremely intensive training. HMB is relatively expensive and the effect would have to be fairly meaningful to justify the cost.

Essential amino acids (EAAs)

As discussed back in Chapter 1, dietary amino acids can be roughly separated into the categories of essential and non-essential (or indispensable and dispensable) depending on whether or not they must be obtained from the diet. There are 8 essential amino acids (EAAs) that cannot be made within the body and so must be obtained by the body from the diet. On average, all high quality proteins contain roughly 40-50% EAAs.

It appears that the EAAs have a number of specific effects in the body that the non-essential amino acids do not. Recent research has shown that only the EAAs are required to stimulate protein synthesis (55); in that study, 13.4 grams of EAAs along with 35 grams of sucrose were given, blood AA levels started to increase within 10 minutes and net protein balance changed from negative to positive although the overall effect was fairly small.

Additional research, mentioned previously, found that there is apparently a dose-response limit in terms of how much EAAs are needed to maximally stimulate muscular protein synthesis (56). Young and old subjects were given EAAs in varying doses from 0 to 20g EAAs (older subjects were given an additional dose of 40g EAAs). In young subjects, 10 g EAAs generated a maximal protein synthetic response; no further increase was seen with the 20g EAA dose.

This amount of EAAs represented roughly 0.13 g EAAs/kg body weight; since whole proteins contain anywhere from 40-50% EAAs (depending on the protein), this would represent 0.26-0.32 g/kg whole protein to generate a maximal response in skeletal muscle. A 100kg lifter would require 13 grams EAAs or roughly 26-32 grams whole protein to maximize muscular protein synthetic response. I'd note again that a maximal muscle protein synthetic response doesn't mean a maximal response in all other tissues and pathways that use AAs; it may be that more protein than this can still have additional physiological effects.

In addition, I'd mention again the study I discussed in Chapter 7 where small amounts of EAAs given in-between solid meals generated an anabolic response without impacting on the response to normal meals (57). Again, differences in total AA intake between groups make drawing conclusions from this study difficult.

As discussed in Chapter 3, there's no indication that stimulating protein synthesis in this fashion leads to gain in skeletal muscle. Consider that eating whole proteins/meals also stimulates protein synthesis in sedentary individuals but that this doesn't lead to significantly increased muscle mass; the body simply adapts by breaking down more protein later in the day. Rather, resistance training in addition to adequate protein intake are required to stimulate gains in muscle mass.

Early research examined the impact of EAAs either by themselves (58,59) or in combination with sucrose (60) around training; both EAAs and EAAs plus carbohydrate stimulated protein synthesis following training. In another study, 6 grams of EAAs were given either by themselves or in a 6% carbohydrate solution during training, muscle fiber breakdown was decreased, cortisol was blunted and insulin levels were higher (61). Recall from Chapter 8 that studies on endurance athletes show the same effect with whole protein added in small amounts to carbohydrate drinks consumed during exercise.

Another study gave subjects 6 grams of EAAs with 35 grams of sucrose either before or after training (62), the pre-training group showed a greater increase in protein synthesis (readers can refer back to Chapter 8 for more discussion of protein timing around training).

Interestingly, as discussed in the nutrient timing chapter, the same was not seen when whey protein was given immediately before training (63), probably due to the relatively slower rate of digestion of whey protein compared to EAAs. It's possible that whey given 30-60 minutes before training would have generated a similar response as it would have had time to digest and release aminos into the system; this has not been tested.

Only one study has examined the long-term impact of EAAs on the adaptations to training. In that study, untrained women were given an extra 18.3 grams of EAAs along with a combined strength training and endurance training program for 6 weeks (64). The study found no improvement in body composition or strength in the supplement group; however treadmill time to exhaustion increased.

Application

For the most part, I see no real benefit to EAA supplements if an athlete is consuming sufficient protein in the first place. As mentioned above, all high quality proteins contain 40-50% EAAs depending on the source. As well, while early studies examined the impact of EAAs on training, subsequent studies have used whole proteins such as whey, whey plus casein, or even milk and found a similar anabolic response to training. Outside of possibly being superior immediately before training when their rapid digestion may make them superior, EAAs would seem to provide no benefit over whole proteins.

Given the high cost and often bitter taste of EAAs powder, it makes more sense to just take 12-15 grams of whole protein instead of 6-8 grams of EAAs powder; the protein can be consumed 30 minutes prior to training to allow adequate time for digestion. A 25g scoop of whey or MPI after training would provide approximately 10-12.5g EAAs.

If an athlete were trying to cut calories to the bone while ensuring sufficient EAA intake, these types of products might have some benefit. Even there, I think whole proteins are a better choice. Consider that a protein powder such as casein or MPI not only provides slower digestion and an anti-catabolic effect, but the dairy calcium may have additional fat loss benefits. EAAs won't help with fullness or provide the anti-catabolism that slowly digesting whole proteins provide.

Finally, as discussed in Chapter 7, it's possible that consuming EAAs with carbs in-between normal meals might be anabolic although there is no indication that this leads to greater lean body mass gains in the long-term. A dose of EAAs of 0.13 g/kg (13 grams for a 100kg athlete) consumed with a small amount of carbohydrates would be expected to give the maximal response.

Creatine monohydrate

Creatine may be the single most studied product in the history of dietary supplements. While not an amino acid itself, creatine is produced in the body from the amino acids arginine, glycine, and methionine. To completely review all of the research on creatine would be unrealistic, due to the sheer volume of it (hundreds of papers); I'll rely heavily on review papers for this reason.

Creatine is found in skeletal muscle in the form of creatine phosphate and its main mode of action is to provide an inorganic phosphate to resynthesize ATP during high intensity activities, typically lasting 10 seconds or less (i.e. sprinting, high intensity weight lifting). However, as more research has been done, it appears that creatine may have benefits for other types of exercise as well. For completeness, I'd mention that creatine has even shown benefit in the treatment of certain disease states but that topic is beyond the scope of this book.

The body normally synthesizes about 1 gram of creatine per day with an additional 1 gram coming in from the diet (65). Studies repeatedly show that muscular stores can be supersaturated with a variety of dosing protocols; creatine loading is discussed below. Conceptually, creatine loading is similar to the idea of carbohydrate loading whereby high intakes of carbohydrates can overfill muscular glycogen stores to improve athletic performance.

Although other forms of creatine have recently become available, the primary form for supplementation is creatine monohydrate (CM). Over the past several decades, a tremendous amount of research has been done on CM and performance.

Creatine supplementation can potentially impact on exercise performance through a number of different mechanisms (65). The most direct is by increasing muscular stores of creatine phosphate which is used to resynthesize ATP during high-intensity activities lasting a few seconds or so. Creatine may also help to buffer changes in muscle acidosis, as well as sparing glycogen utilization during short-duration activities. Creatine doesn't appear to affect aerobic metabolism (66). Thus creatine supplementation would be expected to impact on activities lasting between 10 seconds and 2 minutes with little to no benefit for activities lasting longer than that (the potential impact of creatine on endurance athletes is discussed further below). Creatine also appears to improve recovery between repeated bouts of high intensity activity such as repeated sprints or intervals.

The majority of early studies examined its effects on fairly short-term activity, either weight lifting or sprint performance. Although not all studies found a positive effect, the grand majority did. In one review of 22 studies, the CM group showed an average 8% better gain in maximum strength and a 14% increase in the number of repetitions done with a sub-maximal load (67).

In another review, it was reported that nearly 70% of creatine studies had reported a benefit to high intensity training with none showing a negative effect (68); the other 30% found no effect at all. It reported an average gain in strength and repetitions to failure of 5-15%, an increase in sprint performance of 1-5% and work done during repeated sprint performance of 5-15%.

In addition, creatine has been suggested to increase the gains in muscle mass with training. Studies have routinely shown an increase in lean body mass with creatine supplementation although this is likely to represent increased water retention in the short-term (65). However, over the longer term, CM could potentially impact on muscle mass gains.

These effects might simply be indirect: by allowing an athlete to work with heavier weights or to get more reps with a given weight, gains might be increased.

However, some research has suggested more direct effects of CM on muscle mass including an increased expression of genes involved in muscle growth as well as myosin heavy chain expression (69,70). Creatine might also impact on cell metabolism by altering cell volume (the amount of water in the cell impacts on a number of biological processes including protein synthesis and breakdown).

One study found that 5 days of creatine loading decreased leucine oxidation (a marker of protein breakdown) although it had no effect on protein synthesis (71). Interestingly, this effect was only seen in male subjects; other studies have shown a gender difference in responses to creatine with women typically gaining less body mass than men (72). However, women seem to receive the same performance benefits from creatine supplementation as men (73).

Not all research has found a beneficial effect of CM on protein synthesis and breakdown. One study of CM supplementation examined a host of effects after resistance training and found no impact of CM on either protein synthesis or breakdown (74). The same group found no effect of creatine on protein synthesis or breakdown at rest as well (75). At this point, a direct effect of creatine on muscle growth is still up to debate, although the impact of creatine supplementation on strength and power performance is essentially unarguable.

With one possible exception, CM is likely to be one of the very few must-have sports supplements for any strength or power athlete interested in maximizing either performance or training. CM is inexpensive, readily available, and proven over hundreds of studies to have a beneficial effect under most conditions relevant to those types of athletes. The one major exception to this is weight class athletes for whom CM could cause problems in making weight due to the increased water retention. Even there, CM could be used during training and then stopped several weeks prior to competition to allow any extra water weight to be lost.

While the benefit of CM for strength/power athletes is well established, the role of CM for endurance athletes is more debatable. Based on its mechanism of action, there is little reason to expect CM supplements to greatly benefit endurance events or any event lasting longer than about 3 minutes. For the most part, research has supported this with CM having no real impact on endurance performance (65).

Some early research even suggested that CM might impair running or swimming performance, most likely by increasing body weight (76); sports where body weight is supported such as cycling or rowing sometimes show a performance improvement but this is an inconsistent finding. Studies examining intermittent endurance performance (interval training) have found some benefit, which is in keeping with the mechanism of CM. Given the use of high-intensity/interval training by endurance athletes, CM could still play a role in improving overall training performance and adaptation.

There is other research suggesting that CM can still be useful for endurance athletes although, once again, the data is far from complete. One study of rowers found that CM supplementation had no impact on aerobic metabolism but improved the increase in lactate threshold with training (77). In another study, CM improved anaerobic power output by 18%, with no impact on aerobic metabolism in a group of triathletes (78). CM supplementation for 5 days also decreased markers of muscle inflammation and damage following a 30 km run (79).

So can CM improve endurance performance? Perhaps. CM might be beneficial during periods of interval training, to improve the quality of training and stimulate better gains in adaptation (similar to the effects for strength and power athletes). Additionally, since many endurance events contain high intensity efforts (e.g. a cyclist trying to make a breakaway or climb a hill), CM might have some benefit there as well.

However, any improvements in anaerobic performance have to be balanced against the weight gain that can occur; the increase in body weight could easily overcome any improvements in power output. The data on muscle damage is interesting but preliminary, if CM can improve recovery following extensive endurance training, by limiting muscle damage, this would be another potential benefit to endurance athletes. This is especially true for runners who may cause muscle damage during training due to the high impact nature of their sport (downhill running also causes a great deal of muscle damage). CM could also benefit endurance athletes by increasing glycogen storage (80).

Although research has consistently demonstrated a performance improvement with CM supplementation, some people appear to be non-responders to creatine and there appears to be a specific biological profile that goes with being a creatine responder versus a non-responder (81). Creatine responders tended to have the lowest starting creatine level, a higher proportion of Type II muscle fibers, a larger muscle cross sectional area and a greater amount of lean body mass. Non-responders were the opposite, having higher starting levels of creatine, less Type II fibers, a smaller muscle cross sectional area and less lean body mass.

One current controversy regarding creatine is over a potentially negative impact of caffeine on creatine intake. Two studies, using different performance tests have found that high doses of caffeine (5 mg/kg or 350 mg for a 70 kg athlete) negated the ergogenic effects of creatine (82,83). Creatine appears to reduce muscular relaxation time after contraction (84); this would be important during sprint types of activities. High doses of caffeine appear to negate this effect. Whether lower doses of caffeine have this type of effect is unknown; it's interesting to note that many early studies provided CM with tea (containing small amounts of caffeine) and the benefits were still observed. As well, whether caffeine negates the benefits of CM during others types of activity is currently unknown.

Potential negative effects of creatine supplementation

A number of potential negative effects of creatine have either been studied or theorized and it's important to look at these (65,85). Early creatine products, which tended to have the consistency of sand and mixed poorly, often caused stomach upset in users, especially at high doses. This is less of an issue now as products generally mix and dissolve easily; the small amount of research done on the topic suggests no real effect of creatine on stomach upset.

However, some athletes still have problems with CM and stomach upset especially if they use the higher dose loading schemes. Typically, 20 grams of CM per day for 5 days has been used to load and the sheer amount of CM can cause problems. I'll present two other loading schemes below in the application section that athletes may wish to experiment with if high dose CM intake causes them stomach upset.

Another possible potential negative effect of creatine is the impact of creatine on the kidneys or liver. Research on healthy individuals has found no impact of creatine on either kidney or liver function, although individuals with a preexisting condition could potentially have problems. Creatine supplementation will increase creatinine excretion which can raise alarm if blood work is done, but this is a normal response.

In the early days of creatine use, there were many anecdotal reports of muscle cramping but direct research has not supported this effect. Cramps are more likely due to the high-intensity nature of the training being done and/or electrolyte imbalances. Ensuring adequate water intake during creatine loading is a key factor to avoid such problems. One study of collegiate football players found that CM actually decreased the incidence of cramping and injury (86).

A final potential negative is the increase in body mass with creatine supplementation, although this is most likely caused by an increase in total body water. This increase in mass can range from 1-2.5 kg (2-5 pounds) over the course of the first week of loading; once again, women seem to see a much smaller increase in body mass than men. This has particular relevance for weight class athletes (such as powerlifters, Olympic lifters or MMA athletes) as the increase in body mass from creatine could potentially prevent an athlete from making weight. Additionally, the increase in body weight with creatine may offset performance benefits in some types of sports (especially distance running where the extra body mass requires more energy to move) as discussed above.

Application

As mentioned above, creatine is naturally occurring in meats which contain 4-5 grams of creatine per kilogram (2.2 pounds). A typical daily dose for CM during a standard loading phase may be 20 grams per day; this would require an unrealistic intake of nearly 10 pounds of meat per day.

Almost all of the studies done to date have used CM for supplementation although many other types of creatine products such as creatine serum, effervescent creatine, creatine phosphate or creatine ethyl ester have been brought to market. For the most part there is little to no research on these alternate products and little to no reason to expect them to perform more effectively than inexpensive bulk CM powder.

Regardless of the form of CM used or the type of loading scheme followed, intramuscular creatine levels will eventually reach a maximum level with any excess creatine being excreted in the urine past that point. The only possible benefit of another form of creatine would be to load muscular stores more quickly or possibly with less gastric upset.

The most common dosing pattern has been to take 20 g/day of CM for 5 days and most research has used that approach. However, alternate loading patterns are possible and may be preferred for various reasons. Ten grams of CM for 10 days or even 3 grams per day for a month can both be effective. At the end of this time period, intramuscular creatine stores will be maximized; the only real difference is how long it will take to load.

It appears that insulin is important for creatine uptake and taking creatine with a simple carbohydrate (87) or carbohydrate and protein may increase uptake (88). Adding lipoic acid (1000 mg/day) to creatine with carbohydrates may also increase uptake and storage (89). Creatine uptake is also increased following endurance exercise. Athletes who want to limit their total carbohydrate intake for some reason may want to take their creatine following training (87).

Following loading, many recommend a maintenance dose of 3-6 grams/day to maintain muscular stores (90). Even with zero supplemental creatine intake, muscular stores of creatine remain partially elevated for at least 6 weeks following loading (90).

As mentioned above, creatine occurs naturally in meats, especially red meat. Vegetarians have been found to have lower muscular levels of creatine and appear to get a greater benefit from creatine supplementation (91) as a consequence of starting with lower initial levels. Athletes who eat large amounts of red meat may not get much out of creatine supplementation for this reason: they already have higher levels of creatine stores.

CM is a granular white powder that usually mixes reasonably well in liquids. It can be put into blender drinks and, as mentioned, consuming CM with carbs or carbohydrate and protein improves uptake; so does taking CM after endurance exercise (whether this holds for resistance training is unknown). Due to concern over poor mixability, some athletes have taken their CM by putting the powder directly into their mouth and washing it down with some liquid. This ensures that no unmixed creatine is left in the glass or mixing bowl.

As a final comment, there has been some suggestion that creatine should be cycled, due to a possible impact of creatine supplementation on the creatine transporter or on normal synthesis. While creatine intake appears to downregulate the creatine transporter in animals, this doesn't appear to be the case in humans (92). Additionally, even if CM supplementation downregulates the body's normal production, this should be a non-issue as long as a daily maintenance dose is used; above normal levels of creatine phosphate in the muscles will be maintained.

As mentioned above, CM is perhaps the single most well-studied sports supplement in the history of the industry and would appear to be one of the few supplement "must-haves" especially for strength/power athletes; it may also have some benefit for endurance athletes.

Whether an athlete chooses to do the typical loading phase (20 g/day for 5 days) or one of the longer-term loading approaches (10 g/day for 10 days or even 3-5 g/day for a month) would seem to be irrelevant, at the end of that time period, muscular creatine stores will be maximized. The daily dose of creatine should be split up with at least one dose (3-5 grams) being put into the post-workout carb/protein shake. After loading, a maintenance dose of 3-5 grams/day would be used post-workout on training days, or with any meal on non-training days.

Carnitine

Carnitine is synthesized from the dietary amino acids lysine and methionine; it is found in the diet in meat and dairy products. While carnitine is found in other tissues, it is found primarily in skeletal muscle (93). One of carnitine's primary roles is to transport fatty acids into the mitochondria (the powerhouse of the cell) to be burned for energy and carnitine has long been touted as a fat burner.

As well, the carnitine content of muscle decreases during exercise, especially high intensity activity, and it has been suggested that carnitine supplements may be important to normalize or maintain skeletal muscle levels.

Most early research found that short-term supplementation of even high doses of carnitine (4-6 g/day) failed to increase skeletal muscle levels (94); a 5-hour infusion of carnitine was also ineffective at raising skeletal muscle stores (95). With poor oral absorption (only about 20% of an oral dose is absorbed in the first place), a large body pool, and generally poor uptake of carnitine into skeletal muscle, it would likely take long-term supplementation to have even a small impact on skeletal muscle carnitine stores (94).

In this vein, one study found that carnitine supplementation during six months of endurance training prevented the 10% decrease in skeletal muscle carnitine seen in the placebo group (96). However, it appears that decreases in carnitine levels are fairly unimportant unless a reduction of 25-50% from normal are reached (this can occur due to certain drugs and in certain disease states); a drop in muscle carnitine levels of only 10% is unlikely to have a major impact on skeletal muscle function in the first place (94).

Carnitine has been hypothesized to impact on exercise performance in a number of ways, most of which are relevant to endurance athletes. These include increasing fat burning (sparing muscle glycogen), replacing muscle carnitine which has been redistributed to acylcarnitine (another form of carnitine in the muscle), activating the enzyme pyruvate dehydrogenase (PDH) which may help to prevent acidosis, improving resistance to fatigue and, as mentioned, replacing the carnitine lost with training (96).

Overall, the studies on carnitine and exercise performance are highly variable, with some showing a benefit and others finding no benefit; this is likely related to differences in the subjects studied, the dose and duration of dosing, and the type of exercise test which was administered. This has led researchers to very different conclusions about the usefulness of carnitine as a sports supplement. Some feel that there is evidence for a beneficial use of carnitine (93) for athletes; others have concluded the exact opposite, that carnitine is essentially useless as a supplement (96).

Even in the positive studies of carnitine, where changes in VO2 max. and fat oxidation have been noted, the effect tends to be very small (for example, a few percentage change in fat oxidation or increase in VO2 max.); however these types of small improvements, which may be insignificant in the laboratory, might be valuable to elite athletes looking to maximize performance (94).

As a fat burner/weight loss agent, carnitine has been found to be relatively ineffective unless a diagnosed carnitine deficiency is present.

The generally poor ability of supplemental carnitine to impact on skeletal muscle stores wouldn't predict a major effect of supplementation so the results of the studies done to date are not entirely surprising: if supplementation doesn't increase skeletal muscle levels in the first place, no effect would be expected.

Recent research has found that increasing insulin levels improves carnitine retention (97). In the first part of that study, subjects were given carnitine followed by 4 drinks containing 94 grams of carbohydrate to increase insulin; urinary carnitine excretion was significantly decreased. In the second part, subjects consumed 2 grams of carnitine followed by two drinks consuming 94 grams of carbohydrate 1 and 4 hours after the carnitine; urinary excretion was again decreased. A follow-up study showed that a threshold of increased insulin is required to generate an effect (98); smaller amounts of carbohydrates may be insufficient to increase carnitine uptake.

The same group demonstrated that skeletal muscle carnitine could be raised by 15% with infusion of carnitine and insulin and this impacted acutely on skeletal muscle fuel utilization, decreasing glycogen use and lactate production, most likely by increasing fat utilization (99). However, it would take 100 days of the oral carnitine plus carbohydrate feeding to increase skeletal muscle carnitine levels by even 10% (95). So while it appears that skeletal muscle carnitine levels, and fuel utilization in muscle, might be effectively manipulated by consuming carnitine with highly insulinogenic carbohydrates, it's somewhat questionable how practical this is. An extremely large amount of carbohydrate is necessary and the effect would take a long time to appear.

Of possible importance to strength/power athletes, two recent studies have found that l-carnitine l-tartrate (2-3 g/day for 3 weeks) decreased the incidence of muscle damage with high-repetition weight training (100,101); an even more recent study found similar effects from either 1 or 2 grams (102). The same group showed that supplementation of 2 grams of l-carnitine l-tartrate per day for three weeks caused an increase in the concentration of androgen receptors (103); this might be expected to improve the adaptations seen to resistance training in the long-term.

Application

In bulk, carnitine is relatively inexpensive and an effective dose would appear to be 2-4 grams/day taken for multiple weeks to months. Some studies have found that 2 grams taken immediately before high intensity activity may increase Vo2 max. and decrease lactate accumulation, but this effect is far from universal.

Endurance athletes may wish to experiment with carnitine to see if it affects aerobic or anaerobic performance; strength/power athletes involved in heavy training might find it effective in decreasing soreness and muscle damage. Carnitine is not recommended as a fat burner as it appears to be ineffective in this regard. Vegetarians are potentially at risk for a carnitine deficiency and might benefit from carnitine supplementation.

As noted above, consuming carnitine with high glycemic carbohydrates increases retention in the body, a key to making carnitine even marginally effective. Athletes wishing to experiment with carnitine may want to include it with their pre- or post-workout nutrition (as per the nutrient timing chapter); the increase in insulin from a combination of carbohydrates and an insulinogenic protein should help to ensure greater retention of the l-carnitine. Any benefits from this strategy are likely to be extremely small, taking a long time to actually show up.

Acetyl-l carnitine (ALCAR)

Acetyl l-carnitine, usually abbreviated as ALCAR, is a modified form of carnitine which has a number of effects distinct from carnitine; it has become very popular with athletes and bodybuilders recently.

Carnitine is actually converted to acylcarnitine in skeletal muscle during high intensity activity; however, the main interest in ALCAR has to do with its potential effects in the brain.

A number of studies have identified a profound number of potential health benefits of ALCAR including the inhibition of brain aging (104), cardio- and neuroprotection, and prevention of a normal age related decrease in mitochondrial function (105). ALCAR may also have a role in the treatment of diabetes by increasing mitochondrial function and improving glucose uptake into skeletal muscle (106).

However, little to no research has examined the impact of ALCAR on sports performance and the fascination with ALCAR among healthy athletes is hard to explain or understand. At best, anecdotally, ALCAR seems to improve focus and alertness among people who take it and one study showed an improvement in reflexes during a video game test in healthy individuals (107).

It's been suggested that ALCAR might have beneficial effects on brain function or hormone levels, most likely by affecting dopamine levels in the brain. Most of this work is based on animal research and it's unclear what relevance this has to humans.

One study did show that some women with hypothalamic amenorrhea (a loss of menstrual function, generally related to dysfunction in the hypothalamus that is tied to inadequate energy intake relative to output) benefited from 2 g/day of ALCAR (108); some of the women's cycles normalized with the supplementation. Given the propensity of female athletes, especially those heavily restricting calories or performing excessive amounts of training, to have hormonal problems related to amenorrhea, ALCAR might be worth considering.

Furthermore, there is often a general drop in dopamine signaling during dieting and this would appear to be part of the overall metabolic changes which can occur (e.g. metabolic slowdown, increased hunger, etc.). Increasing dopamine signaling in the brain while dieting may have beneficial effects (readers may wish to refer to my booklet Bromocriptine for a more detailed look at this topic); based on the one study mentioned above, ALCAR might have some benefit in this regard.

Application

ALCAR appears to have many health benefits but any direct impact on exercise performance for either strength or endurance athletes is lacking. Empirically, people have reported greater clarity and focus when using ALCAR in doses of 1-3 g/day. Doses are typically taken first thing in the morning or immediately before training, often with caffeine, tyrosine or other products. Claims of increased focus, training intensity and work capacity are often made; again this has not been tested.

I have seen no reports of changes in body composition, strength or power with ALCAR use. ALCAR may have benefit for older endurance athletes to protect mitochondrial function; combining ALCAR with lipoic acid has even greater benefits, at least in animal models (109). ALCAR is relatively inexpensive in bulk and is probably worth trying but data in healthy human athletes (or non-athletes) is lacking; almost all of it has been done in a variety of pathological conditions or in animal models.

Typically ALCAR is dosed at 1-3 g/day and it's usually suggested that it be taken on an empty stomach; taking it first thing in the morning is often recommended for this reason. As mentioned, ALCAR is often stacked with caffeine, tyrosine and other compounds before intense training to improve focus and work output. ALCAR taken in the evening may keep some people awake due to a slight stimulatory effect.

Carnosine/beta-alanine

Carnosine is synthesized in the body from alanine and histidine; beta-alanine has recently been used to raise carnosine levels in skeletal muscle and is discussed further below. Carnosine appears to be involved in buffering changes in acidosis during exercise (110) although the overall impact in humans is fairly small (111). In one study, skeletal muscle carnosine levels correlated with fatigue resistance in the last part of a 30 second maximal bike test (112). Carnosine levels are elevated in sprinters and rowers but not in marathon runners (113); bodybuilders also show elevated carnosine levels (114). This suggests that the high intensity training done by these types of athletes increases muscular carnosine levels. Only one study has examined the impact of carnosine supplementation on exercise performance, it found no effect (110).

In rats, histidine deficiency lowers skeletal muscle carnosine and supplementation increases it (115), an athlete eating sufficient protein would be unlikely to have such a deficiency. In horses, the combination of beta-alanine and histidine raises muscle carnosine levels (116) and it's been suggested that supplementation of beta-alanine might have similar effects in humans.

Recently, two studies showed that supplementation with beta-alanine can increase skeletal muscle levels of carnosine in humans and this improved maximal performance in a supra-maximal bicycle test (117,118). Doses of 3.2 to 6.4 g of beta-alanine per day were given in 4-8 divided doses of 800 mg each and subjects cycled to exhaustion at 110% of their maximal power output after 4 and 10 weeks of supplementation; the beta-alanine group improved their performance by 12 and 16 seconds at the 4 and 10 week marks. Two other recent studies have suggested that beta-alanine can improve other aspects of endurance performance as well (119,120).

In another study, strength/power athletes were given a combination of 10.5 g creatine and 3.2 g beta-alanine per day in two divided doses or a placebo; the supplement group maintained a higher average volume and intensity during their training and had greater strength gains, along with greater increases in lean body mass and decreases in fat mass (121).

In the most recent study, trained sprinters were supplemented with 4.8 g per day of beta-alanine for 4-5 weeks and a variety of performance tests were done including isokinetic torque production, isometric endurance and a 400m run (122). While beta-alanine raised carnosine levels in both the calf muscles, and improved performance on the isokinetic test, no improvements in running time or isometric endurance were found.

Application

While supplementation of carnosine itself appears to be relatively ineffective in increasing skeletal muscle carnosine content, recent studies have suggested that beta-alanine; a precursor of carnosine can be effective in raising tissue levels. Improvements in various aspects of cycling performance and strength/power have both been shown with supplementation.

For optimal results, 3.2-6.4 g of beta-alanine should be consumed per day in at least 4 divided doses (i.e. 800-1600 mg per dose); at least one month of supplementation is necessary to increase skeletal muscle levels of carnosine although further effects may be seen with longer periods of supplementation. Beta-alanine can cause a slight histamine response, referring to a mildly painful and stinging flushing or burning sensation of the skin. Beta-alanine with creatine may be an even more effective combination for strength/power athletes.

Taurine

Taurine is an inessential amino acid found predominantly in animal source proteins; like all amino acids, it plays a tremendous number of roles in the human body. Unlike other dietary amino acids, it is not involved in the synthesis of skeletal muscle protein. However, it plays a crucial role in skeletal muscle contractility and force production, with any number of pathological muscular states being possibly related to low or depleted taurine levels (123).

Taurine may also reduce excessive adrenal output and the correction of taurine deficiencies lowers high blood pressure (124). Taurine also acts as an insulin mimic and may play a role in the treatment of diabetes (125); it has also been shown to decrease blood triglyceride levels and may decrease the risk of cardiovascular disease (126). However, it's difficult to see how this research applies to otherwise healthy athletes.

In animal models, exercise causes taurine excretion to occur and skeletal muscle taurine levels were found to be decreased in bodybuilders compared to untrained controls (114). Endurance training also appears to decrease skeletal muscle taurine levels (127). This suggests that supplementation is necessary to maintain optimal muscular levels. I am unaware of any direct research on taurine supplements in strength/power athletes.

In one study, taurine supplementation was shown to decrease oxidative stress caused by endurance training (128). An increase in urinary taurine excretion was seen following a marathon (129) and the researchers suggested that this might be related to muscular impairment and/or damage.

Two studies using Red Bull (a drink containing 43 grams carbohydrate, 160 mg caffeine and 2 grams of taurine) have shown changes in performance during endurance exercise: a higher power output and lower sub-maximal oxygen uptake (130,131). A third found improved sub-maximal endurance, pedaling speed and improvements in mental performance (132). Taurine alone or with caffeine might have some benefit for endurance athletes before an important race or intense workouts. A more recent study found that Red Bull slightly increased upper body muscular endurance but had no impact on anaerobic bicycle performance (133).

Empirically, some have used taurine to help with muscle cramps and as a sleep aid and taurine may be beneficial for athletes using a lot of stimulant compounds (such as the ephedrine/caffeine stack or clenbuterol).

Application

Doses of 3-6 grams/day and more are being used in a variety of dosing protocols for different goals. Some have used it around workouts as an insulin mimic/cell volumizing compound while others are using it at bedtime as a sleep/anti-cramp aid. Bulk taurine powder is extremely cheap and may be worth experimenting with. Some have reported stomach upset with high doses (5+ grams/day).

Citrulline malate

Citrulline malate is another nonessential amino acid, related to arginine, that plays a large role in the urea cycle in the liver. Some research has suggested that it may benefit athletes by promoting aerobic metabolism locally in muscle (134); note that this study was in a rather odd finger movement task which has minimal relevance to normal exercise. Another suggested that citrulline helped to decrease blood acidosis and buffer an ammonia increase during exercise (135). While the occasional bodybuilder or anaerobic athlete has reported an effect from citrulline, it would more likely to effect long-duration endurance activities.

In contrast, a recent study found that citrulline supplementation actually hurt endurance performance (136) and much more research needs to be done on citrulline. Another possible effect of citrulline would be in that, since it is converted to arginine, it might provide a better substrate for nitric oxide (NO) production (137); the topic of NO and exercise performance is discussed below under arginine and ornithine supplements.

Application

A typical dose of citrulline would be 3-6 grams per day. Some have suggested that taking a B vitamin complex with the citrulline is required for optimal results; this has not been studied to my knowledge.

Arginine/ornithine/glutamine

Since the 1980's, a variety of amino acids have been promoted as growth hormone (GH) releasers. By way of introduction, GH is a hormone released by the body under a variety of conditions that is important for normal growth. This is especially true if you're talking about GH deficient individuals where replacing their GH levels to normal has profound effects; however, GH appears to play a minimal role at best in the growth of skeletal muscle in healthy adults. GH appears to play a much more significant role in fat loss however (138).

The earliest GH releasing aminos that were promoted to athletes included arginine and ornithine (139), two bitter tasting powders that did appear to increase GH release in fairly large doses, especially when infused. Even there, not all studies found that GH levels were increased in athletes (140,141). More recently, a fairly small dose (2 grams) of glutamine was found to increase GH levels (142).

It's questionable how great of an impact these small pulses of GH had and most studies have found no anabolic effect of any of the GH releasers; few scientists recommend them as having any sort of anabolic effect (143). Related to endurance athletes, two studies have examined the impact of arginine aspartame supplementation and found no impact on performance (144,145).

However, such compounds could conceivably have an impact on fat loss. GH is one of several lipolytic (fat mobilizing) stimuli although it is not a primary stimulator of fat loss. Research has shown that even physiologic pulses of GH can stimulate lipolysis, although the effect takes 2-2.5 hours to show up (146). Additionally, inhibition of the normal nighttime GH pulse decreases fat breakdown on the subsequent day (147).

Athletes trying to lose fat might benefit from taking a few grams of glutamine (cheaper and better tasting than high doses of arginine/ornithine) at bedtime to ensure a proper GH release and optimal fat mobilization the next day. What real-world impact this might have is unknown and has not been directly tested. Recall from above that glutamine may have other important roles in terms of immune function.

In addition to the supposed GH releasing effect, arginine has generated interest of late due to its conversion to a compound called nitric oxide (NO). NO is a profoundly important signaling molecule in the body and is involved in blood flow in many tissues along with many other effects. Viagra, which promotes erections, does so by inhibiting the enzyme that breaks down NO in the penis, promoting blood flow to that area.

Therapeutically, arginine may be of great benefit for individuals suffering from a variety of disease states related to heart disease (148). At least one researcher has suggested that arginine might have an ergogenic effect by increasing blood flow to muscles (and thus increasing amino acid delivery) if added to other amino acids; this has not been tested in humans (148).

Arginine is the primary compound in most of the NO products currently available and while claims for increased growth have been made, I am unaware of any research to support this; many users report an improved pump during training although the real-world significance of this in terms of muscular adaptation is unknown.

From a practical standpoint, it's questionable how much of an impact arginine actually has on NO in healthy individuals. Raising NO via arginine supplementation or infusion certainly appears to have some benefit in individuals with heart disease and diabetes (148); however, it doesn't appear that arginine is effective in raising NO or improving exercise performance in otherwise healthy individuals (149, 150).

As a final note, I should mention that insulin increases NO production (and blood flow) to many tissues including skeletal muscle; eating carbohydrates with protein prior to will raise NO levels as well (151). Following the pre-workout nutrient recommendations from Chapter 8 should be more than sufficient to increase NO release and blood flow during training.

Application

None of the supposed GH releasers are recommended for anabolic purposes although they might play a small role during dieting by increasing fat mobilization. The rank taste and high doses required for both arginine and ornithine make them poor choices; encapsulation is always a possibility but this increases the cost and a large number of pills will have to be taken to achieve an effective dose.

Taking several two to three grams of glutamine at bedtime (to ensure a proper nighttime GH pulse) or two to three hours before training might increase lipolysis. Whether or not this will impact on overall fat loss on a properly set up diet and training program is unknown.

While arginine might have the potential to increase skeletal muscle blood flow due to the stimulation of NO release, there is no real indication that this has any benefit in terms of a training response. Consuming a small amount of high glycemic index carbs with protein prior to training increases insulin; not only will this raise NO levels, it will provide nutrients to the skeletal muscle and promote a much greater anabolic effect than any NO precursor could provide.

N-acetyl cysteine (NAC)

I mentioned the importance of cysteine and its role in promoting the production of glutathione (one of the body's inherent antioxidant defense systems) in the last chapter during the discussion of whey protein; the high cysteine content of whey has been shown to increase the body's store of glutathione and this may confer health benefits (152).

From a performance standpoint, it appears that a high cysteine intake may improve anaerobic performance (153); of interest to dieters, one study found that individuals with low blood glutamine levels were at greater risk for body protein loss; supplementation of 400 mg per day of n-acetyl-cysteine prevented this loss from occurring (154).

Application

Athletes consuming whey protein on a daily basis should already be obtaining a significant amount of cysteine. However, athletes who find that they lose lean body mass while dieting may wish to experiment with n-acetyl-cysteine. A dosage of 400 mg per day in two divided doses appears to be effective.

4-hydroxyisoleucine

4-hydroxyisoleucine is an amino acid isolated from fenugreek seeds, it has been used in traditional medicine to lower blood glucose levels with research showing that it increases insulin release (155), lowering blood glucose in diabetic individuals (156).

Of some relevance to athletes, a recent study found that it could increase the rate of glycogen synthesis following high-intensity endurance depleting endurance exercise (157). In that study, subjects received either 1.8 g/kg of carbohydrate or the same amount of carbohydrate with 2 mg/kg of 4-hydroxyisoleucine; muscle glycogen storage was roughly 63% faster for the supplemented group. In contrast, a recent study found no effect on glycogen storage following 5 hours of lower intensity cycling (158).

Application

As discussed in the chapter on nutrient timing, athletes who perform two or more workouts per day often need to be concerned with optimal resynthesis of glycogen between workouts. In addition to providing sufficient carbohydrate, along with protein, athletes who need to maximize glycogen storage might consider adding 4-hydroxyisoleucine to their post-workout drink following high-intensity training. As mentioned above, 2 mg/kg body weight should be added to the post-workout drink. For a 100 kg (220 lb) athlete, that would be 2.0 grams of 4-hydroxyisoleucine.

Bovine Colostrum

Colostrum is a specialized protein found in mother's milk that contains a number of bioactive peptides important for gut health (159). In the first 24-48 hours of life, those proteins ensure that the baby's gut finishes developing; this contributes to overall health as well. Interest in colostrum for both health and athletic benefits have been examined and bovine colostrum (taken from cow's milk) has been brought to market.

Colostrum lies somewhere between a protein powder and a protein based supplement. Containing a variety of proteins found in mother's milk, colostrum wouldn't typically be used in the same fashion as a protein powder, but neither is it an amino acid per se.

As with many supplements, the research on colostrum is highly varied. Early research suggested that colostrum could raise blood levels of insulin-like growth factor 1 (IGF-1, 160) but not all studies find this to be true (161). Given that IGF-1 is not absorbed orally to a significant degree (160), any effect of colostrum on IGF-1 levels would have to be mediated through another mechanism.

Some research has found a benefit of colostrum supplementation on athletic performance. This includes increases in anaerobic peak power (161), increases in recovery (but not performance) from endurance training (162), increased cycling time trial performance (163) and increased sprint speed in field hockey players (164). In one study, supplementation of colostrum was found to improve body composition changes when coupled with training (165).

In other studies, colostrum failed to improve rowing performance (166); one study suggested that colostrum increased skin thickness and that accounted for the gains in arm circumference (167). Two recent studies compared 10 grams of bovine colostrum with 10 grams of whey protein over 8 weeks of cycling training including 5 days of consecutive high intensity interval training, the colostrum group showed greater improvements during a 40 km time trial along with a decreased incidence of upper respiratory tract illness (168, 169). The exact mechanism by which bovine colostrum may improve performance is currently unknown.

Application

At the recommended dose, bovine colostrum is extremely expensive. Although commercial products often list 6 grams/day as an effective dose, most studies have used 20-60 grams/day and generally compared it to a concentrated whey protein. Two recent studies have used a more realistic dose of 10 grams per day over 8 weeks although this would still represent a significant daily cost. For the most part, unless an athlete had their training and other dietary aspects absolutely perfect and money was no object, colostrum simply isn't a cost effective product.

Summary

In writing this chapter, I've tried to touch on a majority of popularly available protein or amino acid based supplements.

Clearly there are products with a good deal of human research available. Creatine, beta-alanine, and the BCAAs have been studied extensively for their role in athletic performance. Other products have less supporting research but could be beneficial under certain conditions. Still others are clearly ineffective.

I'd simply reiterate that prior to adding any supplement to their dietary regimen, athletes should ensure proper overall nutrition and training first. All of the supplements in the world won't make up for insufficient protein, insufficient dietary energy or a poorly planned training program.

Most find that, after a proper diet and training program are in order, supplements add only a small amount. However, for competitive athletes seeking those final few percentage points to their performance, adding specific supplements to a proper training and nutrition program may be worthwhile.

13

Putting it All Together

In the previous 12 chapters, I've covered a tremendous amount of information regarding protein. In this chapter I want to get into practical applications of all of this information for different types of athletes. As I've discussed throughout this book, it's impossible to give any single set of recommendations that can apply to all athletes equally.

Differences in the type and amount of training, along with the specific goals of a given athlete, all serve to impact on what might constitute an ideal amount of total protein to be consumed each day (1). Those same differences would impact on the amounts and types of nutrients that might ideally be consumed around training, both to support training and for optimal recovery afterwards.

Little to no research has examined the impact of different whole food proteins and how they might apply to different types of sports. However, emerging research has compared the impact of different types of protein powders (e.g. soy versus milk protein) on training adaptations, primarily muscle growth following resistance training. As discussed in Chapter 12, the various protein or amino acid based supplements have relatively more or less application to endurance and strength/power sports respectively.

In order to address these topics in an applied way, I first want to divide sports into one of two primary categories, with an additional sub-category. The main determinants of which category a given sport falls into are the duration of the event during competition, as well as the predominant type of training done.

In addition to the physiological demands of different sports, an additional issue that impacts on optimal protein nutrition has to do with varying goals of the athlete. I'm going to simplify these into maintaining current body mass (while improving performance), losing body fat, and gaining body mass (generally muscle mass). I'll address how each goal impacts on total protein requirements along with the ideal types and timing of protein during the day and around training.

To finish the chapter, I want to examine issues specific to female athletes and address the issue of vegetarianism as it pertains to protein nutrition for athletes.

Sport categories

Sports can be subdivided into a number of different categories depending on the physiological requirements of competition and the specifics of their training. Throughout this book, I've divided sports and training into two primary categories which are endurance and strength/power. In this chapter, I'm also going to introduce a third sub-category under strength/power sports. That category is the physique sports: bodybuilding, fitness and figure. While they share many aspects in common with strength/power sports, they are distinct enough in terms of their training and physiological requirements to be considered separately from a nutritional point of view.

Pure endurance sports include activities such as cycling, running, cross country skiing, rowing and others. Such sports are typified by extended durations and relatively continuous activity. The intensity of such events is typically sub-maximal, although many such sports require bursts of high intensity activity during competition as well (e.g. a cyclist climbing a hill or covering a break, a distance runner doing the same).

The majority of training for endurance sports is typically done for extended periods at sub-maximal intensities; weight training typically plays a much smaller overall role and endurance athletes generally aren't pursuing increased muscle mass as a primary goal. Weight training is often done for injury prevention or performance reasons and many endurance athletes perform some amount of weight training during their off-season or early preparation phase, often when normal training is limited by bad weather. Even there, large increases in muscle mass are rarely an explicit goal of pure endurance type sports.

The strength/power category simply includes everything else. Any sport where a greater proportion of the training is done under anaerobic conditions, or where strength/power training (in the weight room or out) is a primary focus would be included in the strength/power category. This category includes everything from pure strength/power events such as the throws, power and Olympic lifting and short sprints, to medium duration events such as the 400 or 800m sprints. Team sports such as football, rugby, basketball, hockey, etc. would also be included in this category. Since some type of metabolic conditioning or endurance work is also often done, athletes in this category often have to cover the protein requirements for both types of activities.

Practically speaking, any sport where relatively large amounts of strength and/or power training is done should follow the daily protein recommendations for strength/power athletes; sports which don't rely so heavily on resistance training or where increases in muscle mass aren't the goal and where training primarily entails high volumes of lower intensity exercise should use the recommendations for endurance athletes.

Although they share many aspects in common with strength/power sports in terms of their training (a majority of which is done in the weight room), the physique sports are unique enough to be placed into their own category. For the most part, the physique activities are judged on appearance and are not, strictly speaking, performance based. One exception to this is the fitness routine in fitness competitions. The differences between the physique sports and the more performance oriented strength/power sports have implications for all aspects of diet, which I'll address below.

Although the above sporting classifications should be used to determine total daily protein intake (i.e. endurance athletes should use the endurance recommendations and strength/power athletes should use the strength/power recommendations), one question that arises is which set of guidelines for around training nutrition should be used.

In general, I think it's better to think in terms of the type of workout being performed (rather than the sport category per se) to be used when determining which recommendations to use for around workout training. A strength/power athlete performing an endurance type of conditioning workout would be better served by utilizing the recommendations for around endurance nutrition (although lower carbohydrate values would most likely still be used) as the requirement for protein simply will not be as high as when a strength/power workout is being performed. An endurance athlete engaging in resistance training would be better served using the strength/power recommendations (although the lower protein values would generally be used since muscle mass gains are generally not the goal).

Sporting goals

Obviously, the goal of all sports training is improved performance. Whether this means throwing further, lifting a heavier weight, running, cycling or swimming faster or what have you, the fundamental goal of the training process is to improve competition performance. For the physique sports, increased muscle mass or fat loss could be considered an explicit performance goal as well.

While the topic of optimal sports nutrition is beyond the scope of this book, hopefully the issue of proper protein intake has been sufficiently covered. Ensuring optimal protein intake has many potential benefits from promoting optimal adaptations to training, to limiting protein breakdown during endurance training, to helping to stave off overtraining.

In addition to promoting optimal performance improvements, a common goal among athletes has to do with changes in body mass. Losing body fat, gaining muscle mass (and sometimes simply total body mass) are both common goals for athletes. As noted above, these may be the primary goals of athletes in the physique sports.

The first topic I want to address in this chapter is how an athlete can go about applying the information from this book when their goals are essentially maintenance of current body composition along with promoting optimal adaptations to training and improvement in performance. I'll recap the guidelines for total protein intake as well as recommendations for nutrient timing around training, discuss how to make both whole food and protein powder choices, and finally make a few comments about the use of supplements.

Having established the maintenance situation as a baseline, I then want to discuss how an athlete's diet should change in terms of total protein intake, protein choices, etc. when their goals include fat loss or muscle gain as explicit goals.

Maintenance: Introduction

When athletes reach an appropriate level of leanness and/or muscularity for their sport, the goal of training and nutrition then becomes to generate maximal adaptation to their training for improved performance. A huge part of this, of course, is a properly designed training program which is far beyond the scope of this book.

Nutritionally, promoting optimal training adaptations encompasses many topics which are also outside the scope of this book. Micronutrients (vitamins and minerals), fluid, and sufficient energy (from carbohydrates and fat) to cover both the energy cost of training along with the energy needed to synthesize new proteins are all required.

Ensuring adequate protein intake, along with the above, is an important aspect of an optimal diet for athletes. While it's true that many athletes already consume sufficient (or even excessive) amounts of protein, it's also true that many consume insufficient amounts to meet their goals.

In addition, current research strongly supports the idea that the timing of nutrients around training may be just as important in terms of improving performance, recovery, and adaptation as the total amount consumed per day. This is currently a huge area of research.

Dietary supplements, while all too often used to make up for deficiencies or cover up mistakes in the diet, can often be useful or have small additional effects when added to a proper diet and training program.

When changes in body composition aren't an explicit goal, athletes use a variety of different measurements of performance to gauge whether or not their training and dietary program is effective. Strength or metabolic tests or simply on-field performance can all be used to determine if an athlete is progressing or not. As well, even when maintenance of body composition is the goal, changes in body composition can indicate if a supposed maintenance intake is appropriate or not. As mentioned above, athletes in the physique sports may be using changes in body composition as their primary or only indicator.

If body fat is accumulating, an excessive number of calories are being consumed and a slight reduction in intake (or increase in activity) should occur; if body weight/body fat is decreasing (and this isn't an explicit goal), calories should to be increased to compensate. Assuming that protein intake is set at appropriate levels, changes in caloric intake should come from adjustments in carbohydrates or fat. Finally, if body weight/body fat is stable but performance improvements aren't occurring, a slightly increased caloric intake (to ensure sufficient nutrients and energy for optimal adaptation) may be needed.

It's worth noting that certain subgroups of athletes, notably female runners, dancers and gymnasts along with male wrestlers and boxers are known to chronically restrict food intake to keep their body weight down and this often limits their ability to train intensely; this can cause a variety of health problems as well, especially for females. I'll address some issues specific to female athletes later in this chapter.

Maintenance: Specific recommendations

Athletes who are at maintenance should be following the protein recommendations set earlier in this book and shown again in Table 1.

Table 1: Daily protein recommendations

Type of athlete	Habitual		Dieting	
	g/kg	g/lb	g/kg	g/lb
Male strength/power	2.5-3.0	1.1-1.4	3.0-3.3	1.4-1.5
Female strength/power	2.4-2.6	1.1-1.2	2.6-3.0	1.2-1.4
Male endurance	1.7-2.0	0.7-0.9	2.0-2.2	0.9-1.0
Female endurance	1.3-1.6	0.6 -0.7	1.6-1.9	0.75-0.9

Athletes who are at maintenance will be using the values in the "Habitual" column which I'd note, covers a fairly large rang of intakes. Given the impact of energy intake on nitrogen balance, with calories at maintenance a protein intake anywhere within those ranges should be fine.

For endurance athletes, as either the duration or intensity of training goes up, using the values nearer the higher end of the protein recommendations is going to be most appropriate. During periods of lower training volumes or intensities, protein intake can be decreased if desired. I'd mention that many endurance athletes report "feeling" better with higher protein intakes. There may be no need to deliberately lower protein intake during periods of lower intensity training for this reason.

For strength/power athletes, the same basic comments apply: a protein intake anywhere in the above range should be appropriate. Those who are explicitly trying to avoid gains in lean body mass may wish to set protein intake nearer the lower levels while athletes still wishing to make small gains in lean body mass (which technically isn't maintenance) may wish to consume protein at the higher levels.

In the physique sports, training approaches vary significantly from very high volumes of training to lower volume/high-intensity approaches. In general, athletes in these activities tend to default to diets that are higher in protein and lower in carbohydrate and often fat. Assuming that training intensity and quality don't suffer, setting protein intake at the high end of the recommendations with a reduction in carbohydrate or fat intake makes sense for this group.

In Table 2 on the next page, I've calculated out the daily protein intake that athletes of different body sizes would require based on the intake recommendations given in Table 1.

Table 2: Daily Protein intake for athletes of different body mass

	g/kg	1.1	1.3	1.5	1.8	2.0	2.2	2.4	2.6	2.9	3.1	3.3
	g/lb	0.5	0.6	0.7	0.8	0.9	1.0	1.1	1.2	1.3	1.4	1.5
Kg	Lbs											
45	100	50	59	69	79	89	99	109	119	129	139	149
50	110	55	66	77	88	99	110	121	132	143	154	165
55	121	60	73	85	97	109	121	133	145	157	169	182
60	132	66	79	92	106	119	132	145	158	172	185	198
65	143	72	86	100	114	129	143	157	172	186	200	214
70	154	77	92	108	123	139	154	169	185	200	216	231
75	165	82	99	116	132	148	165	182	198	214	231	248
80	176	88	106	123	141	158	176	194	211	229	246	264
85	187	94	112	131	150	168	187	206	224	243	262	280
90	198	99	119	139	158	178	198	218	238	257	277	297
95	209	104	125	146	167	188	209	230	251	272	293	314
100	220	110	132	154	176	198	220	242	264	286	308	330
105	231	116	139	162	185	208	231	254	277	300	323	346
110	242	121	145	169	194	218	242	266	290	315	339	363
115	253	126	152	177	202	228	253	278	304	329	354	380
120	264	132	158	185	211	238	264	290	317	343	370	396
125	275	138	165	192	220	248	275	302	330	358	385	412

The left column shows body weight in both kilograms and pounds, the top row shows protein intake in both g/kg and g/lb values. Athletes who are heavier than 125 kg (275 pounds) or lighter than 45 kg (100 pounds) will have to break out the calculator.

Given that many athletes don't train every day, a question that often comes up is whether or not protein should be kept at the recommended levels on off-days when training is not being performed. In general the answer is yes as adaptation to training is an ongoing process. For example, protein synthesis following resistance training can remain elevated for 24-48 hours (2,3). I'm unaware of similar data on endurance training.

In any case, providing sufficient building blocks and energy on off-days is generally a good idea to ensure maximal adaptations to training. At the same time, it might be argued that protein requirements would be slightly lower on non-training days since the immediate support of training adaptations (or limiting protein breakdown during training) isn't necessary.

One possible solution would be to use the lower values for protein recommendations on non-training days and the higher value on training days. So strength/power athletes might aim for 2.5 g/kg (1.1 g/lb) on non-training days and a full 3.0 g/kg (1.4 g/lb) on training days; endurance athletes would use 1.7 g/kg (0.7 g/lb) on non-training days and a full 2.0 g/kg (0.9 g/lb) on training days. The difference in protein intake between training and non-training days would generally be made up by the nutrients consumed around training.

Maintenance: Non-training days

On non-training days, athletes should simply divide up their daily protein intake across however many meals they are consuming. As discussed in Chapter 7, eating anywhere from every 3-5 hours is the most reasonable approach and this will typically yield anywhere from 4-6 meals/day; the day's protein would be divided fairly evenly across each meal.

In terms of protein choices, at maintenance calories, protein quality won't matter that much, especially on non-training days. One fairly unlikely exception would be an athlete choosing a single low quality protein source to fulfill their daily intake; even there, the sheer quantity of protein ingested should make up for any issues with quality.

I recommend athletes get a variety of proteins on a day-to-day basis. As discussed in detail in Chapter 10, all proteins tend to have their own pros and cons. Many athletes, especially those in the physique sports, tend to fixate on one or two protein sources. Not only does this mean that they may miss out on potential benefits of other sources, but they may be compounding potential negatives inherent to the sources that they do consume (e.g. excessive mercury intake from a reliance on certain types of seafood).

I'd like to emphasize again the important benefits I think that lean red meat contribute to the diet, especially for females athletes in terms of iron status. While red meat needn't be eaten daily, I do feel it should be eaten several times per week to ensure sufficient iron intake (in addition to the other important micronutrients found in meat). I'd also reemphasize the benefits of dairy proteins in terms of both their high quality and potential positive impact on body composition.

As I mentioned in Chapter 10 there is some logical reason to think that consuming varied proteins at the same meal might provide some benefits; any limitations of one protein could be counteracted by the presence of another protein. Dairy such as milk, yogurt or cheese can usually be added easily to other protein meals (e.g. glass of milk, cup of yogurt, cheese on top of a burger or salad), beans and other legumes add fiber, carbohydrates, calories and protein to the meal; many other possibilities can be dreamt up. Different types of animal proteins (i.e. chicken and fish or red meat) could be eaten at a single meal.

On non-training days, protein powders probably aren't necessary except for reasons of convenience or practicality. They may be helpful for athletes who have very high protein requirements but who have difficulty consuming sufficient amounts of whole food. Athletes who work full time jobs or who are in school may need to mix protein drinks (recall that carbohydrates, fat and fiber should ideally be present) to consume between main meals; others like to use protein powders to make dessert type items. Athletes with a poor appetite often find that drinking some portion of their daily calories is easier than eating solid foods and protein powders can be used to mix high calorie blender drinks.

With regards to supplements, whether or not specific supplements are taken on non-training days will depend entirely on the nature of the supplement. Obviously the products that require chronic use such as creatine or beta-alanine to maintain tissue levels would be consumed on non-training days. Compounds used solely around training or to support some aspect of training wouldn't need to be consumed on non-training days.

Maintenance: Training days

Depending on the nature of the sport and the athlete, athletes may be training anywhere from once every other day to twice or more six days per week at the highest level of sports training. Generally speaking, workouts tend to fall into one of two major categories which are weight training and metabolic work (endurance or conditioning training).

Weight training is an extremely broad category and can range from hypertrophy work (typically 6-12 rep sets) for a physique athlete or performance athlete who needed to gain muscle mass to a low repetition (1-5 rep sets) workout geared towards maximal strength and/or power. Strength-endurance training utilizing higher repetitions and short rest periods are used by both strongmen and performance athletes who need to improve local muscular endurance. Combined workouts including both low repetition work for strength/power and higher rep work for hypertrophy; structural balance or general conditioning are often done during these workouts as well.

Each workout has its own specific physiological demands in terms of the fuel used, amount of glycogen depleted, amount of muscle damage done, and the amount of protein synthesis that is stimulated. Thus the requirements for protein and calories to support training can vary profoundly between the different types of resistance training. In general, athletes emphasizing lower repetitions and strength won't have the requirements for either calories or protein around training as compared to athletes looking for more hypertrophy or even strength-endurance (where a great deal of muscle glycogen may be depleted).

Metabolic or endurance work also takes a tremendous number of forms. True endurance sport training can range from short (30-60 minutes) recovery workouts at very low intensities, to extensive workouts (2-8 hours) at low to moderate intensities to high intensity interval or time trial work ranging from 1 to 20 minutes or longer (the last type of training typically makes up a relatively small percentage of the total training time). Again, caloric costs can vary significantly between workouts although endurance athletes, in general, have higher energy requirements than strength/power athletes (4). A moderate caloric burn of even 600 calories/hour sustained for 3-4 hours equates to 1800-2400 calories burned in addition to maintenance levels.

Strength/power athletes may be doing body weight or medicine ball circuits, complexes in the weight room or various types of interval work for conditioning, "endurance" or work capacity training. Generally speaking strength/power athletes don't and shouldn't perform a lot of low-intensity steady state endurance type work (running distance, etc.) as this tends to impair the strength/power adaptations that they are seeking (5).

Of course, other types of training are often done. Pure skill or technical training is perhaps one of the most typical; generally speaking, this doesn't have the high-energy cost or recovery needs of the other types of training but the high levels of concentration required can be mentally draining.

As mentioned above, athletes have essentially two options as to how protein intake might vary on training versus non-training days. Either protein intake will remain constant every day (training or otherwise) or the different "ends" of the range would be used on non-training days versus training days with slightly larger intakes being consumed on training days relative to non-training days.

In either case, some proportion of the day's total protein intake will be placed around training, meaning that it will need to be subtracted from the other meals of the day. Simply, having determined how much protein will be consumed in total on the training day based on Tables 1 and 2 above, the amount being consumed around training would be subtracted from that and the remainder distributed through the day's meals.

As well, between the energy cost of training itself, along with energy needed for tissue and protein synthesis, calories will generally be higher on training days versus non-training days, with a majority of the difference coming from carbohydrates consumed before, during, and/or after training.

Because of the differences in nutrient absorption and utilization during exercise, as compared to pre- or post-workout, readers may recall that the during workout recommendations were expressed in absolute terms (grams per hour) rather than relative to body weight. Because of this, I want to discuss them separately in this chapter before moving onto pre- and post-workout intake. In Table 3, I've recapped the during workout recommendations from Chapter 8.

Table 3: During workout nutrition recommendations

	Protein	**Carbohydrate**
Endurance	8-15 g/hour whey protein	30-60 g/hour dextrose alone OR 45-70 g/hour dextrose plus fructose
Strength/power	12-15g/hour whey protein	30-45 g/hour dextrose or sucrose

Note: Other carbohydrates such as maltodextrin, maltose or sucrose can be used in place of dextrose if desired or better tolerated by the athlete. As well, soy could be used in place of whey.

A typical strength/power workout might last one to two hours, so an athlete might be consuming 12-30 grams of protein total along with 30-90 grams of carbohydrate. An endurance type of workout might represent anywhere from 1-6 hours for a pure endurance athlete while a strength/power athlete might only perform 1 hour of metabolic conditioning. So while the strength/power athlete might only consume 15 grams of protein with 30 grams of carbohydrate, a cyclist on the bike for 4 hours might be consuming 32-60 grams of protein and up to 240 or more grams of carbohydrate during their workout.

Clearly the amount of protein consumed during training can vary significantly because of this, with the endurance athlete consuming far more total protein during training due to their workouts generally being much longer in duration. And while it seems that the endurance athlete might actually be consuming a majority of their daily protein during the workout, I'd note that an endurance athlete who is training 4-6 hours/day and sleeping 8 has a fairly limited time to consume the remainder of their food; proportionally more of their intake *should* be consumed during training since it represents a much larger part of their total day. In any case, the amount of protein consumed during training will simply be subtracted from the daily totals.

Having determined the amount of protein that will be consumed during training, the final consideration for training days becomes that of pre- and post-workout nutrition. Below I've reprinted the pre- and post-workout recommendations from Chapter 8 in Table 4 and 5. I've also added a new Table 6 which simply adds up the pre- and post-workout recommendations to show the total amount of protein and carbs that would be consumed around training.

Table 4: Immediate Pre-workout nutrition recommendations

	Protein	**Carbohydrate**
Endurance	0.15-0.25 g/kg	1.0 g/kg
Strength/power	0.3-0.4 g/kg*	0.3-0.4 g/kg

* Half of this value, or 0.15-0.25 g/kg of EAAs could be used instead

Table 5: Post workout nutrition recommendations

	Protein	**Carbohydrate**
Endurance	0.15-0.35 g/kg	1-1.85 g/kg
Strength/power	0.3-0.5 g/kg	0.3-1.5 g/kg

Table 6: Total pre/post workout nutrition recommendations

	Protein	**Carbohydrate**
Endurance	0.3-0.6 g/kg	2.0-2.85 g/kg
Strength/power	0.6-0.9 g/kg	0.6-1.9 g/kg

Again, Table 6 simply represents the amounts shown in Tables 4 and 5 added together and would represent the total amount of protein consumed pre- and post-training. This would be added to the amount being consumed during training (from Table 3 above), and that total amount would be subtracted from the day's overall totals; the remainder would simply be divided up throughout the day's whole food meals.

Table 7 on the next page shows the total amount of protein in grams that will be consumed based on the totals from Table 6 above.

As with Table 1, the left hand column shows body weight in kilograms and pounds, the top row shows total protein intake recommendations in both g/kg and g/lb. So a 70 kg (154 pound) endurance athlete consuming 0.6 g/kg of protein around training would be consuming 42 grams of protein, this would be added to the amount of protein being consumed during training, and the total amount of protein consumed around training would be subtracted from the daily protein intake.

A 100 kg (220 pound) strength/power athlete consuming 0.9 g/kg of protein around training would be consuming 90 grams of protein total around training which would be first added to any protein consumed during training, and the total amount of protein consumed around training would be subtracted from his daily totals.

Table 7: Total protein intake around training

	g/kg	0.3	0.4	0.5	0.6	0.7	0.8	0.9	1.0
	g/lb	0.14	0.18	0.23	0.27	0.32	0.36	0.41	0.45
Kg	Lbs								
45	99	13.5	18	22.5	27	31.5	36	40.5	45
50	110	15	20	25	30	35	40	45	50
55	121	16.5	22	27.5	33	38.5	44	49.5	55
60	132	18	24	30	36	42	48	54	60
65	143	19.5	26	32.5	39	45.5	52	58.5	65
70	154	21	28	35	42	49	56	63	70
75	165	22.5	30	37.5	45	52.5	60	67.5	75
80	176	24	32	40	48	56	64	72	80
85	187	25.5	34	42.5	51	59.5	68	76.5	85
90	198	27	36	45	54	63	72	81	90
95	209	28.5	38	47.5	57	66.5	76	85.5	95
100	220	30	40	50	60	70	80	90	100
105	231	31.5	42	52.5	63	73.5	84	94.5	105
110	242	33	44	55	66	77	88	99	110
115	253	34.5	46	57.5	69	80.5	92	103.5	115
120	264	36	48	60	72	84	96	108	120
125	275	37.5	50	62.5	75	87.5	100	112.5	125

To show you how to make the calculations, let's assume a male strength/power athlete who weighs 100 kg (220 pounds) and who is consuming the full 3 g/kg of protein per day from Table 1; he will be consuming a total of 300 grams of protein per day. On a particular training day, let's assume that he performs a strength/power session which lasts one hour and he'll be consuming 15 grams of protein during the workout. He prefers to consume the highest levels of protein before and after training so he'll be consuming an additional 0.9 g/kg of protein or 90 grams around training.

In total around training, he's consuming 105 grams of protein (90 grams pre- and post- plus 15 grams during his workout). Subtracted from his daily intake of 300 grams of protein, that leaves 195 grams of protein which would be divided relatively evenly through his remaining daily meals. If he's consuming 4 whole food meals, that would be approximately 48 grams of protein per meal; if he's consuming 6 meals, that's 32.5 grams of protein per meal.

In terms of protein choices around training, as described in Chapter 8, using a fast protein before and during training appears to be optimal. Whey would be the first choice for pre- and during workout intake due to its rapid digestion; the high BCAA and EAA content is also a plus. For athletes allergic to dairy proteins or who are vegetarians, soy protein is another fast protein which would be appropriate before and during training. I should mention that some athletes prefer to forego whole proteins during training, preferring to consume BCAA mixtures. As discussed in Chapter 12, this may conceivably cause fatigue during training due to increased production of ammonia, a problem that doesn't occur when whole proteins such as whey are consumed.

Post-workout, emerging data suggests that a slow protein such as casein or a mixed fast and slow protein such as milk protein isolate appears to be superior. Alternately, whey protein could be mixed in milk, or milk itself could be consumed. It would generally be difficult for a strength/power athlete to get sufficient protein post-workout with milk without consuming a tremendous amount. Due to their lower protein requirements post-workout, an endurance athlete might find milk to be effective as both a source of carbohydrate and fat; the fluid content also facilitates rehydration.

Many athletes simply prefer to eat a whole food meal after training; little data exists and the final determinant will be whether or not the athlete is actually hungry enough to eat a solid meal following training. Some are but many are not, making liquids a preferred choice. Finally, various supplements can be consumed around training as part of the pre-, during- or post-workout drink.

Losing body fat: Introduction

For most sports, ensuring sufficient leanness without sacrificing performance is an important goal and many athletes have to actively diet to reach those levels. Under most circumstances, excess body fat (above some optimal level, which varies by sport) is simply dead weight that is a negative to performance. At the same time, many athletes become so obsessed with losing body fat that they sacrifice performance via impaired training capacity or lost muscle mass in an attempt to reach unrealistic and unnecessary levels of leanness.

Sports where athletes are projecting themselves against gravity such as running, road cycling, the jumping events, gymnastics, etc. generally benefit from lowered body fat levels from a performance standpoint. Overall body mass is often fairly low in such sports as well.

Athletes in weight class sports such as wrestling and the lighter powerlifting and Olympic lifting classes also typically benefit from lowered body fat as this allows the athlete to maximize the amount of muscle mass that they can carry at a given body weight. Weight class athletes often use a combination of fat loss and dehydration to make it into a lower weight class as well.

Bodybuilding, figure and fitness are all judged on both muscularity and leanness (fitness has the added component of the fitness routine) and fat loss, often to extreme levels, is a common goal. It's not an over-exaggeration to say that many individuals in those sports are professional dieters. Maximizing fat loss while minimizing muscle loss has been a long-standing goal in the physique sports.

The challenge with fat loss is finding ways to optimally lose body fat without sacrificing training capacity, performance or muscle mass. Tangentially, under some very rare circumstances, an athlete may wish to lose unnecessary muscle mass but this tends to be the exception rather than the rule.

Losing body fat: Specific recommendations

At a basic level, losing body fat requires a prolonged imbalance between caloric expenditure and intake; the athlete must be burning more calories than they are consuming. Under those circumstances, the body will draw on stored body fat to make up the deficit, causing fat loss over time. However, the body can also use protein for fuel and this can cause muscle and performance loss; this becomes especially problematic as athletes try to reach lower levels of body fat.

In practice, creating a deficit either means reducing total caloric intake, increasing activity, or some combination of the two. Looking at food intake, athletes have four primary food groups from which to reduce their food intake: protein, carbohydrate, fat and alcohol. As alcohol provides nothing from a performance or nutritional standpoint, reducing its intake can be one simple way of reducing total caloric intake. Unless athletes are consuming a large amount of alcohol on a daily basis, caloric restrictions will realistically have to come from protein, carbohydrates or fats.

In general, unless intake is already excessive, caloric reductions should not come from protein. Rather, as discussed in Chapter 4, protein requirements go up while dieting and dietary protein intake may need to be increased somewhat when calories are being restricted. With alcohol and protein eliminated as possibilities, this means that reductions have to come from carbohydrates or fat. There are advantages and disadvantages to reducing either carbohydrates or fat but they are beyond the scope of this book.

As mentioned above, protein intake should be increased slightly when calories are reduced. Table 1 above shows recommended protein intakes while dieting for different groups and Table 2 can be used to recalculate protein intake while dieting.

As discussed in Chapter 8, meal frequency appears to play a much larger role in terms of nitrogen balance and lean body mass maintenance during periods of lowered caloric intake, more so when protein intake is low. With the high protein recommendations being made in this book, it's debatable how much of an impact meal frequency will actually have.

As discussed in that chapter, someone consuming 50 grams of protein in a sitting will have amino acids being released into the bloodstream for 5 hours or more, depending on the source. In any case, ensuring that the body has a sufficient supply of amino acids throughout the day by spacing meals evenly is the safest and most time-tested and proven dietary strategy to limit lean body mass losses.

In addition, slower digesting proteins, including whole foods and casein, appear to have a distinct benefit during periods of dieting due to the anti-catabolic effect of maintaining constant levels of amino acids in the bloodstream. As discussed in Chapter 11, whey may also have benefits in terms of muscle sparing due to its high cysteine content.

Milk protein isolate (MPI) may be an ideal protein powder for dieting for these reasons. In addition to containing both whey and casein, MPI has a reputation for keeping dieters full longer than other powders, helping to allay the hunger that can derail a diet. As well, as discussed in both the chapter on whole food proteins and protein powders, the consumption of dairy proteins on a diet has been shown to increase the loss of body fat, making their inclusion in a fat loss diet that much more important.

One strategy worth considering is to consume a small amount of casein or MPI with a small amount of fat first thing in the morning and again at bedtime, around normal meals. This would keep a sustained level of amino acids in the bloodstream for many hours, helping to limit muscle loss. The combination of protein and fat also helps to blunt hunger.

Beyond those specific protein recommendations, the same basic idea as for maintenance intake holds: consuming a variety of proteins while dieting helps to ensure that any negatives are minimized while positives of each protein are maximized. It's even more common for athletes to default to one or two protein sources while dieting and this can lead to their missing out on important nutrients.

As one very specific example, many dieters concerned with fat intake will eliminate red meat, which compromises iron and zinc intake. Given the prevalence of very low-fat red meat (and the importance of consuming some fat while dieting), this seems unnecessary. As long as total caloric intake is controlled, all protein sources can be safely included on a diet. I'd note that, as discussed in Chapter 10, physique athletes may need to remove dairy in the last few days of their contest preparation if they are having issues with water retention. Beyond that, I see no need to eliminate any particular category of proteins while dieting.

On non-training days, as with maintenance, the day's protein intake should simply be divided across the day's meals. On training days, the typical question revolves around the importance or benefit of nutrition around training. Many athletes reduce or eliminate nutrition around training either to save calories or in an attempt to increase fat mobilization.

As discussed in Chapter 8, consuming nutrients around training doesn't appear to hamper fat loss in the first place. More importantly, the maintenance of proper training intensity on a diet is far more important than any (arguably) small impact on fuel mobilization. An inability to train intensely and effectively on a diet tends to lead to performance and muscle loss; ensuring sufficient nutrients around training goes a long way towards eliminating this problem.

Rather than eliminating nutrients around training, I prefer athletes to reduce food intake at other meals throughout the day. That is, carbohydrate or fat reductions should be made at the meals not consumed around training. However, to allow for sufficient food intake at those meals, the amounts of nutrients consumed around training may need to be cut back slightly (especially the carbohydrate intake).

Athletes may want to use the lower recommendations for carbohydrate intake before, during and after training. This strategy should help to promote optimal training intensity and recovery while still allowing sufficient calories so that the other meals of the day aren't too small.

However, with the exception of low-intensity aerobic work, which doesn't require a lot of around workout nutrition in the first place, I believe that proper pre-, during and post-workout nutrition should be maintained to some degree even while dieting.

Although I'd generally refer readers back to Chapter 12 for a discussion of protein based supplements while dieting, I do want to mention a current practice being used by some athletes (primarily bodybuilders) which is to sip on small amounts of branched chain amino acids (BCAAs) during the day while dieting. This idea comes from the single study in wrestlers where massive intakes of BCAAs exerted a small protein sparing effect. Recall from that chapter that the wrestlers were receiving roughly 1.1 g/kg (0.5 g/lb) of protein, roughly 1/3rd of what this book recommends while dieting.

As discussed in that chapter, my gut feeling is that, given the high intake recommendations for protein given in this book, coupled with an intake of either whey/casein or MPI (all containing high amounts of BCAAs), that extra BCAAs are probably unnecessary. An athlete consuming 3.3 g/kg (1.5 g/lb) of protein is already getting roughly 0.66 g/kg of BCAAs (based on the fact that most proteins are roughly 20% BCAAs), more if they are consuming whey protein. I have trouble seeing how anything but massive extra doses would have any further benefit. Athletes who feel that they are prone to muscle loss on a diet may still wish to consider this strategy.

N-acetyl-cysteine (NAC) may also be a useful supplement for those individuals since it appears to spare muscle mass while dieting for individuals with low cysteine levels; note that athletes consuming whey protein before and during training as recommended throughout this book are probably already getting sufficient amounts of cysteine in their diets.

Gaining mass: Introduction

Another very common goal among athletes is an increase in muscle mass. This can be done as an end unto itself (i.e. for the physique sports) or to provide a basis for either increased strength or power production for performance based activities.

Additionally, some sports simply require a large overall body mass even if some of the extra weight gained is body fat. Football linemen are one example that come to mind; super-heavyweight power and Olympic lifters often carry relatively large amounts of body fat in addition to their muscle mass.

It would be somewhat rare, although not unheard of, for endurance athletes to want to increase muscle mass; generally, if this is done at all it is done during the early part of the season (or during the winter months when outdoor training tends to be curtailed).

Gaining mass: Specific recommendations

As with the discussion of fat loss above, fundamentally gaining body mass is fairly simple: an athlete needs to consume more calories than they expend and body weight in some proportion of fat and muscle will typically be gained. For those athletes unconcerned with the quality of weight gained, simply "eating more" may be a sufficient recommendation. However, for those athletes seeking to gain predominantly muscle mass, a little more detail is needed.

It should go without saying that gaining skeletal muscle mass requires a proper training program. Although muscle mass can't be built out of nothing in terms of energy and building blocks, the body simply won't see any "need" to build new muscle mass without an appropriate stimulus; that stimulus is a proper weight training program, the details of which are beyond the scope of this book.

However, once a proper training program is in place, providing both sufficient building blocks in the form of protein/amino acids and energy intake is key. Any athlete seeking maximal gains in muscle mass should increase total protein intake to the high end of the range; this includes endurance athletes who, for whatever reason, wanted to gain muscle mass. Tables 1 and 2 should be used to determine daily protein intakes.

I should mention that, empirically, anabolic steroid users seem to benefit from intakes even higher than those recommended in this book, intakes of 4.4 g/kg (2 g/lb) or even higher are often consumed in order to maximize the effects of the drugs.

I want to emphasize that, as discussed in Chapter 4, once protein requirements are met, increases in total energy intake tends to have a far greater impact on lean body mass gains than simply consuming more protein. Ensuring sufficient carbohydrate and fat intake on a day-to-day basis, in addition to adequate protein intake, is more important than just consuming more and more dietary protein. However, massive energy surpluses are generally not necessary as they only promote excess gains in body fat.

In terms of meal frequency, once again there is limited data and it doesn't appear that meal frequency or spacing plays a major role in the body's retention of nitrogen at above maintenance calories. I'd mention again that it does appear that eating too frequently could actually be detrimental to the goal of gaining muscle mass although this topic needs to be researched in more detail. As long as an athlete is consuming sufficient amounts of protein as part of a mixed meal every 3-5 hours, the body will have more than sufficient amino acids to support growth.

There is limited data at best on whether any specific type of whole food protein is superior to any other in terms of gaining lean body mass. As mentioned above, at above maintenance calories, protein quality becomes much less relevant. Athletes seeking gains in muscle mass should follow the same recommendations given for maintenance above and consume a variety of proteins throughout the day.

The same basic idea holds for protein powders. Athletes who find it difficult to consume the recommended protein intakes from whole foods, or who have scheduling difficulties that make consumption of meals a problem (i.e. work/travel) may find protein drinks of benefit. Whey, casein or MPI is still the first choice but other proteins may also be useful from either a cost or convenience standpoint.

In terms of around workout nutrition, athletes seeking maximal muscle mass gains would want to consume nutrients at the high end of the strength/power recommendations around their resistance workouts to maximize the anabolic effect of training. Ensuring sufficient protein and carbohydrates around training has been shown repeatedly to promote protein synthesis, inhibit protein breakdown and promote gains in muscle mass. Once again, a fast protein before and during training and a slow or mixed fast/slow protein after training appear to be ideal. Whey before and during training and casein, MPI or whey mixed in milk afterwards would be appropriate. Many athletes prefer to eat a solid meal following training and clearly this is effective as well.

Athletes seeking maximal muscle mass often concern themselves with the overnight fast and the fact that, during this time, the body is going without nutrients. This tends be a special concern of athletes who perform their resistance training in the evening; this can be a major period of time during which protein synthesis is elevated but is occurring without the availability of nutrients. One potential solution is to consume a slow protein (along with carbohydrate, fiber and fat) at bedtime to provide nutrients and energy through at least part of the night. A whole food meal or a drink containing casein or MPI along with other nutrients at bedtime should provide protein and energy during at least part of the fasting period.

Some athletes go even further than this, waking up in the middle of the night to consume another protein drink. I'm mixed as to the benefits of this strategy. On the one hand, anything that interrupts the normal sleep cycle (and hence recovery) should generally be avoided and setting an alarm to wake up in the middle of the night may interrupt the recovery that occurs during this period. At the same time, athletes who naturally wake up (e.g. to use the bathroom) in the middle of the night could easily consume a premixed drink (or a glass of milk) to maintain nutrient levels throughout the night.

A final strategy discussed in Chapter 7 has to do with the possible benefit of essential amino acid (EAA) supplements and mass gains. As discussed in that chapter, one study found that consuming small amounts of EAAs between whole food meals had an additional effect on protein synthesis, although a lack of a good control group draws the results into question. As well, whether or not this strategy actually yields greater gains in lean body mass compared to eating whole food meals is currently unknown.

As far as supplements go, creatine is in the must-have category when the goal is increased muscle mass. Between its established effects on training capacity and potential effects on gene expression, coupled with low-cost, there's no reason not to consume it. Emerging research on beta-alanine also makes it a worthwhile consideration; by allowing athletes to maintain a higher workload, better gains may occur.

Female athletes

As mentioned above, female athletes, because of their lighter weight (combined frequently with concerns about body weight) tend to restrict calories to a greater degree than men.

This is especially prevalent in sports such as gymnastics, ballet, and other sports where maintaining an extremely low body weight (for either performance or aesthetic reasons) is an integral part of the sport.

Additionally, in my experience, female athletes often tend to restrict protein intake, either out of concern for the protein itself, or in a misguided attempt to eliminate dietary fat from their diets. Along with this can come a poor intake of several specific micronutrients including calcium, iron and zinc (6). Ensuring sufficient protein intake from varied sources, including dairy and red meat, is an important aspect of both general health and optimal athletic performance for females.

Conversely, between generally smaller caloric requirements, coupled with an often self-imposed restriction of total energy intake, athletes who fill up their daily caloric requirements with protein can have difficulties consuming sufficient energy from carbohydrate and fat to support training. Of course, ensuring adequate total caloric intake in the first place should go a long way towards eliminating this problem.

As discussed back in Chapter 4, there is some indication that females may need less protein than men although most of the research to date has been done on endurance training. Although some have suggested no difference in males and female strength/power athletes in terms of protein requirements, based on fundamental physiology alone (i.e. differences in the rate of muscle mass increase), female athletes can probably get by with slightly less protein than men on a day to day basis. As well, due to their generally higher body fat levels, estimates based on total body weight may already be over-estimating female protein needs. Hence my recommendations for slightly reduced values.

Once again, Table 1 above includes recommended intakes for female athletes under both habitual and dieting conditions. Table 2 can be used to determine daily protein intakes.

Physiologically, there seems to be no reason for females to follow a different meal frequency than male athletes. However, practically this isn't always the case as discussed back in Chapter 7. Female athletes with low total daily caloric intakes often find a high meal frequency leads to extremely small meals.

For example, a light female athlete consuming 1500 calories per day and trying to eat 6 small meals per day would only be eating 250 calories/meal. Split into 4 more realistic meals per day, this would yield 375 calories/meal. Clearly, as total caloric intake goes up, the issue of high meal frequencies leading to disappointingly small meals becomes less of an issue.

Outside of the issues discussed above, notably poor calcium, iron and zinc intake, I see no reason for women to choose different protein sources than men. I'd simply encourage female athletes to make skim- or low-fat dairy and low-fat red meat a regular part of their diet to ensure that micronutrient intake is adequate.

Similarly, I see no fundamental reason for females to choose different protein powders than men if they are used. I'd only mention that both whey and casein can serve as calcium sources, as they are on par with whole dairy foods in terms of their calcium content. Due to the phytoestrogen content, soy protein is still an area of much debate; I'd refer female readers back to Chapter 10 for a more detailed discussion upon which to base their choice.

Out of a misguided desire to reduce caloric intake, female athletes often eliminate around workout nutrition in an attempt to limit what they perceive as "unnecessary" caloric intake. For the same reasons discussed under fat loss above, this is a mistake. Ensuring sufficient nutrients around training to maintain proper training intensity and support recovery is just as crucial for female athletes as for males. And since the nutrient recommendations are scaled to body weight, a lighter female will be consuming less absolute calories anyhow.

Even there, because of generally lowered caloric (and potentially protein) requirements, females may wish to consume nearer the low end of the around workout recommendations. This should be done only as long as training intensity can be maintained and recovery and progress are both occurring. As with dieting, I'd encourage athletes to consume sufficient calories around intense training even if it means reducing intake at other meals of the day.

With regards to supplements, once again there's no reason for female athletes to have different requirements or recommendations compared to males. About the only comment I'd make has to do with creatine: females who are particularly weight conscious should be aware of the slight weight gain that creatine often causes due to water retention.

Vegetarianism

Vegetarian style eating patterns are sometimes practiced by athletes and, due to the elimination of certain categories of foods, there are specific nutritional issues that vegetarian athletes need to be aware of (7).

In order to discuss some issues related to vegetarianism and athletics, I need to define some terms first as there are many different "types" of vegetarianism practiced. At the least extreme, many vegetarians will eat all animal source foods with the exception of red meat. So fish, chicken, eggs, and dairy are acceptable. Strictly speaking this isn't truly a vegetarian diet but this is not an uncommon eating pattern. Lacto-ovo vegetarians will consume milk and eggs but no fish, chicken or red meat. Some variants allow fish but no land animals (chicken or red meat). At the most extreme is strict veganism which disallows any animal source foods including those that come from animals such as eggs, milk or honey.

Without getting mired in various moral or other arguments for or against vegetarianism, I want to point out that there are both pros and cons to a vegetarian lifestyle. Diets high in vegetables and fruits are consistent with good health and athletic performance; athletes should consume ample amounts of those foods whether their diet contains meat or not.

However, vegetarian athletes can have a few issues to contend with, again depending on the extremity of their eating patterns. A general truism is that the more food groups that are eliminated from the diet, the more difficult it becomes to obtain optimal amounts of all nutrients. This would be true for a meat-eating athlete who did not consume vegetables or fruit just as much as a vegetarian athlete who omits meat from their diet; both groups are at risk for insufficient nutrient intake due to their avoidance of an entire food group.

For the most part, obtaining sufficient total amounts of quality protein is no problem for vegetarians who eat milk and eggs; those that allow some type of animal flesh (chicken/fish) should have absolutely no difficulty either. Recall from Chapter 5 that adding even small amounts of higher quality proteins to vegetarian protein sources tends to enhance the overall protein value of the meal. As long as some form of concentrated protein source is being consumed, total protein intake is generally a non-issue.

Strict vegans often find that, due to the relatively non-concentrated nature of most vegetable protein sources, they have trouble obtaining sufficient protein without consuming an enormous amount of food. Achieving the amounts of protein recommended in this book from purely vegetarian sources would be extremely difficult without having to consume massive numbers of calories. The few vegetarian protein powders (soy, hemp, rice, pea) can facilitate sufficient protein intake for vegans.

And although this seems to contradict the previous paragraph, the high bulk nature of vegetarian types of diets often causes spontaneous reductions in caloric intake and vegetarians are often found to have lower body weight levels than meat-eaters; this might make a more vegetarian based lifestyle attractive for athletes who need to lose weight. However, for athletes with extremely high caloric requirements, vegetarian diets may make it difficult to consume sufficient calories to cover training needs.

Outside of the potential issue of difficulties achieving adequate total protein intake (again depending on the extremity of food restriction), vegetarians also have to deal with the potential for micronutrient imbalances as a consequence of eliminating specific foods. Deficiencies in zinc, B12, calcium and vitamin D are the primary issues. While total iron intake may be adequate in non-meat eaters, the lower absorption of non-heme iron in non-animal sources can predispose vegetarians towards iron insufficiency.

Recall from Chapter 10 that a factor present in both red meat and chicken improves iron absorption from non-heme sources. Vegetarians who consume chicken should try to consume other vegetarian iron sources at the same meal to maximize iron absorption. Cooking in a cast iron skillet and consuming vitamin C with iron containing meals also improves absorption.

Clearly the types of proteins consumed by vegetarians will depend on the type of vegetarianism practiced, I'd still encourage vegetarian athletes to obtain protein from as many varied sources as possible within the limits of their specific type of vegetarianism. Vegan athletes will almost assuredly have difficulty obtaining the levels of protein recommended in this book without supplementation with protein powder.

In terms of protein powders, again what is allowed will depend on the type of vegetarianism practiced. Lacto-ovo vegetarians can consume whey, casein or MPI and essentially follow every recommendation made in this book in terms of overall use of protein powders or around workout nutrition.

Vegetarian athletes who avoid dairy products could use soy protein as a fast protein for before and during workout nutrition. I am unaware of any slow vegetarian proteins but soy could be mixed with other nutrients to slow its digestion. Vegans will have to rely on the few purely vegetable based protein powders for any supplementation.

Creatine is typically low in vegetarians with some research suggesting that vegetarians respond better to creatine than meat-eaters. Supplementation by vegetarian strength/power athletes is highly recommended.

Appendix 1: Protein intake Tables

In this appendix, I've reprinted the various tables for recommended protein intakes from previous chapters. Table 1 recaps the overall daily protein recommendations from Chapter 4

Table 1: Daily protein recommendations

Type of athlete	Habitual		Dieting	
	g/kg	g/lb	g/kg	g/lb
Male strength/power	2.5-3.0	1.1-1.4	3.0-3.3	1.4-1.5
Female strength/power	2.4-2.6	1.1-1.2	2.6-3.0	1.2-1.4
Male endurance	1.7-2.0	0.7-0.9	2.0-2.2	0.9-1.0
Female endurance	1.3-1.6	0.6 -0.7	1.63-1.92	0.75-0.9

Table 2 simply calculates the protein recommendations from Table 1 above for individuals of different bodyweight. The left column shows body weight in both kilograms and pounds, the top row shows protein intake in both g/kg and g/lb values. Athletes who are heavier than 125 kg (275 pounds) or lighter than 45 kg (100 pounds) will have to break out the calculator.

Table 2: Daily Protein intake for athletes of different body mass

	g/kg	1.1	1.3	1.5	1.8	2.0	2.2	2.4	2.6	2.9	3.1	3.3
	g/lb	0.5	0.6	0.7	0.8	0.9	1.0	1.1	1.2	1.3	1.4	1.5
Kg	Lbs											
45	100	50	59	69	79	89	99	109	119	129	139	149
50	110	55	66	77	88	99	110	121	132	143	154	165
55	121	60	73	85	97	109	121	133	145	157	169	182
60	132	66	79	92	106	119	132	145	158	172	185	198
65	143	72	86	100	114	129	143	157	172	186	200	214
70	154	77	92	108	123	139	154	169	185	200	216	231
75	165	82	99	116	132	148	165	182	198	214	231	248
80	176	88	106	123	141	158	176	194	211	229	246	264
85	187	94	112	131	150	168	187	206	224	243	262	280
90	198	99	119	139	158	178	198	218	238	257	277	297
95	209	104	125	146	167	188	209	230	251	272	293	314
100	220	110	132	154	176	198	220	242	264	286	308	330
105	231	116	139	162	185	208	231	254	277	300	323	346
110	242	121	145	169	194	218	242	266	290	315	339	363
115	253	126	152	177	202	228	253	278	304	329	354	380
120	264	132	158	185	211	238	264	290	317	343	370	396
125	275	138	165	192	220	248	275	302	330	358	385	412

Tables 3 through 6 reprint the recommendations for around workout nutrition from Chapters 8 and 13.

Since nutrient intake recommendations during training are based on the amount that can be absorbed from the gut rather than scaled to bodyweight, they are presented below in Table 3 separate from the pre- and post-workout nutrient recommendations.

Table 3: During workout nutrition recommendations

	Protein	Carbohydrate
Endurance	8-15 g/hour whey protein	30-60 g/hour dextrose alone OR
		45-70 g/hour dextrose + fructose
Strength/power	12-15g/hour whey protein	30-45 g/hour dextrose or sucrose

Note: Other carbohydrates such as maltodextrin, maltose or sucrose can be used in place of dextrose if desired or better tolerated by the athlete. As well, soy could be used in place of whey.

Tables 4 and 5 recap the pre- and post-workout recommendations for nutrient intake. Table 6 lists the total values found by adding up the values from Tables 4 and 5.

Table 4: Immediate Pre-workout nutrition recommendations

	Protein	Carbohydrate
Endurance	0.15-0.25 g/kg	1.0 g/kg
Strength/power	0.3-0.4 g/kg*	0.3-0.4 g/kg

* Half of this value, or 0.15-0.25 g/kg of EAAs could be used instead

Table 5: Post workout nutrition recommendations

	Protein	Carbohydrate
Endurance	0.15-0.35 g/kg	1-1.85 g/kg
Strength/power	0.3-0.5 g/kg	0.3-1.5 g/kg

Table 6: Total pre/post workout nutrition recommendations

	Protein	Carbohydrate
Endurance	0.3-0.6 g/kg	2.0-2.85 g/kg
Strength/power	0.6-0.9 g/kg	0.6-1.9 g/kg

Table 7 on the next page shows the total amount of protein in grams that will be consumed based on the totals from Table 6 above.

Table 7: Total protein intake around training

	g/kg	0.3	0.4	0.5	0.6	0.7	0.8	0.9	1.0
	g/lb	0.14	0.18	0.23	0.27	0.32	0.36	0.41	0.45
Kg	Lbs								
45	99	13.5	18	22.5	27	31.5	36	40.5	45
50	110	15	20	25	30	35	40	45	50
55	121	16.5	22	27.5	33	38.5	44	49.5	55
60	132	18	24	30	36	42	48	54	60
65	143	19.5	26	32.5	39	45.5	52	58.5	65
70	154	21	28	35	42	49	56	63	70
75	165	22.5	30	37.5	45	52.5	60	67.5	75
80	176	24	32	40	48	56	64	72	80
85	187	25.5	34	42.5	51	59.5	68	76.5	85
90	198	27	36	45	54	63	72	81	90
95	209	28.5	38	47.5	57	66.5	76	85.5	95
100	220	30	40	50	60	70	80	90	100
105	231	31.5	42	52.5	63	73.5	84	94.5	105
110	242	33	44	55	66	77	88	99	110
115	253	34.5	46	57.5	69	80.5	92	103.5	115
120	264	36	48	60	72	84	96	108	120
125	275	37.5	50	62.5	75	87.5	100	112.5	125

The left hand column shows body weight in kilograms and pounds, the top row shows total protein intake recommendations in both g/kg and g/lb. So a 70 kg (154 pound) endurance athlete consuming 0.6 g/kg of protein around training would be consuming 42 grams of protein, this would be added to the amount of protein being consumed during training, and the total amount of protein consumed around training would be subtracted from the daily protein intake.

Appendix 2: Determining Protein Cost

As mentioned in Chapters 10 and 11, one aspect of rating proteins has to do with cost. Unfortunately, it's impossible to give any sort of ranking of proteins by cost that will be of any use; prices change routinely and costs can vary by locale. Rather, I'd like to show readers how to make their own chart to compare varying protein sources.

From a strictly financial point of view (and again, this ignores issues of quality, accessory nutrients, taste, etc.), probably the best way to rank proteins is in terms of cost per gram of protein; this tends to be more valuable than just looking at cost per serving. This is important, for example, because some protein powders (notably some of the vegetarian powders such as rice, hemp and pea; whole egg protein powder also contains significantly less protein per serving due to the presence of dietary fat) aren't very concentrated so the actual cost per gram of protein is much higher. So one protein may only cost 1.5 cents per gram of protein while another costs 4 cents per gram of protein.

Once again, please note that the actual value of a protein may only be partially related to its protein content. The presence of micronutrients like iron and zinc (red meat) or speed of digestion/absorption (whey vs. casein) may be of equal or greater importance depending on the specific situation. Readers should refer back to Chapter 10 and 11 on whole food protein and protein powders respectively for details about individual protein sources.

You will need two numbers to make the calculation of cost per gram of protein. The first is the cost of the protein in your local currency (I'll be using cents in the examples below). The second is the total amount of protein present in grams. Then you simply divide the cost by the total amount of protein in grams, this will give you the cost per gram of protein.

For example, a typical can of tuna fish costs about 50 cents and contains 32 grams of protein. 50 cents divided by 32 grams of protein = 1.8 cents per gram of protein. A cup of low fat yogurt also costs about 50 cents and contains 8 grams of protein (along with 12 or so grams of carbohydrate). This gives it a cost of 50 cents/8 grams of protein = 6.25 cents/gram of protein.

Let's say that we have red meat of some sort that costs $4.50 per pound (one pound is 16 oz). First we convert the total cost of 4.50 to 450 cents. Next we have to determine the total amount of protein present; animal protein typically contains about 7-8 grams of protein per oz so 16 oz yields 112-128 grams of protein, I'll split the middle and use 120 grams of protein. So that's 450 cents/120 grams protein = 3.75 cents/gram.

So from lowest to highest, using these three examples, tuna comes in at 1.8 cents per gram of protein, red meat at 3.75 cents/gram and yogurt at 6.25 cents per gram of protein. Does that make tuna superior? Looking only at the protein content, perhaps. But that has to be weighed against the zinc and iron content of the red meat or the quality/calcium content of the yogurt.

The same basic calculation can be done for protein powders. Looking at one of the major commercial online supplement houses, a 5-pound container of ON Gold Standard whey protein costs $39.00 or 3900 cents. The container has 77 servings of protein, each of which contains 24 grams of protein so the container has 77 * 24 = 1,848 grams of protein. 3900 cents / 1848 grams of protein = 2.11 cents/gram protein.

Milk protein isolate currently retails for between 7.12-7.49 per pound (depending on how much you buy) which is 712 to 749 cents per pound, I'll split the middle and call it 725 cents per pound and split the middle. Each pound contains 15 servings of protein at 26 grams per serving which gives 15 servings * 26 grams protein per serving which is 390 grams of protein. 725 cents / 390 grams of protein = 1.85 cents/gram protein. Hopefully you get the idea of how to do the calculation.

Just using those examples, I can make a chart ranking the proteins from least to most expensive. Again, other aspects of the protein such as the presence of absence of other micronutrients, the rate of absorption or application (i.e. whey protein versus casein) should also be taken into account when deciding which protein is best for a given application. Table 1 below summarizes the values from the examples I gave above.

Table 1: Comparison of Prices for Different Protein Sources

Food	Total protein (g)	Cost (cents)	Cost/gram protein
Tuna fish	32	50	1.8 cents
Milk protein isolate	390	650	1.85 cents
Whey protein	1848	2700	2.11 cents
Red meat	120	450	3.75 cents
Yogurt	8	50	6.75 cents

As I mentioned, many supplemental protein powders are now more economical than some food proteins as the above chart indicates, at least in the US. For overseas athletes, shipping and other costs often drive the price of protein powders much higher and make them much less economical than whole food. Please note again that foods are likely to contain nutrients not found in protein powder and nutritional science is only beginning to determine what nutrients may or may not be involved in either optimal human health or athletic performance and muscle growth.

References

References cited for Chapter 1: Definitions and Basic Background

1. "Advanced nutrition and human metabolism, 2nd ed." James L. Groff, Sareen S. Gropper, Sara M. Hunt. West Publishing, 1995
2. S.A. Laidlaw and J.D. Kopple "Newer concepts of the indispensable amino acids" Am J Clin Nutr (1987) 46: 593-605.
3. Moughan PJ. Et. al. Food-derived bioactive peptides influence gut function. Int J Sports Nutr Exerc Metab (2007) 17: S5-S22.
4. Phillips SM et. al. A critical examination of dietary protein requirements, benefits, and excesses in athletes. Int J Sports Nutr Exerc Metab (2007) 17: S58-S76.

References cited for Chapter 2: Protein Digestion

1. Ten Have, GA. et. al. Absorbtion kinetics of amino acids, peptides, and intact proteins. Int J Sports Nutr Exerc Metab (2007) 17: S23-S36.
2. "Advanced nutrition and human metabolism, 2nd ed." James L. Groff, Sareen S. Gropper, Sara M. Hunt. West Publishing, 1995.
3. Silk, D.B.A. et. al. Protein digestion and amino acid and peptide absorption. Proceedings of the Nutrition Society (1985) 44: 63-72.
4. Benevenga, NJ et. al. Role of protein synthesis in amino acid catabolism. J Nutr (1993) 123:332-336.
5. Jungas, RL et. al. Quantitative analysis of amino acid oxidation and related gluconeogenesis in humans. Physiological Reviews (1992) 72: 419-448.
6. Meguid, MM et. al. Leucine kinetics at graded leucine intakes in young men. Am J Clin Nutr (1986) 43: 770-780.
7. Meguid, MM et. al. Valine kinetics at graded valine kinetics in young men. Am J Clin Nutr (1986) 43: 781-786.
8. Meredith, CN et. al. Lysine kinetics at graded lysine intakes in young men. Am J Clin Nutr (1986) 43: 787-794.
9. Zhao, Xi-he et. al. Threonine kinetics at graded threonine intakes in young men. Am J Clin Nutr (1986) 43: 795-802.
10. Motil, KJ et. al. Whole body leucine and lysine metabolism: Response to dietary protein intake in young men. Am J Physiol (1981) 240: E712-E721.
11. Lemon P. Is increased dietary protein necessary or beneficial for individuals with a physically active lifestyle? Nutrition Reviews (1996) 54: S169-S175.
12. Millward, DJ. The endocrine response to dietary protein: the anabolic drive on growth. In Milk Proteins: Nutritional, clinical, functional and technological aspects. C.A. Barth, E. Schlimme (Eds.) Springer Verlag, 1989.
13. Millward, DJ and Rivers, JPW. The need for indispensable amino acids: the concept of the anabolic drive. Diabetes/Metabolism Rev (1989) 5:191-211.
14. Gelfand, RA et. al. Removal of infused amino acids by splanchnic and leg tisues in humans. Am J Physiol (1986) 250: E407-E413.
15. Ashley, DV et. al. Plasma amino acid responses in humans to evening meals of differing nutritional composition. Am J Clin Nutr (1982) 36: 143-153.
16. Wahren, J et. al. Effect of protein ingestion on splanchnic and leg metabolism in normal man and in patients with diabetes mellitus. J Clin Invest (1976) 57: 990-995.
17. Fern, E.B. et. al. Effects of exaggerated amino acid and protein supply in man Experientia (1991) 47: 168-172.

18. Wagenmakers, AJ. Protein and amino acid metabolism in human muscle. Skeletal Muscle Metabolism in Exercise and Diabetes. ed. Richter et. al. Plenum Press: New York, 1998.
19. Waterlow, JC. Protein turnover with special reference to man. Q J Exp Phys (1984) 69: 409-438.
20. Furst, P. Intracellular muscle free amino acids - their measurement and function. Proc Nutr Soc (1983) 42: 451-462.
21. Scriver, CR et. al. Normal plasma amino acid value in adults: The influence of some common physiological variables. Metabolism (1985) 34: 868-873.
22. Moriarty, KJ et. al. Relative nutritional value of whole protein, hydrolysed protein and free amino acids in man. Gut (1985) 26: 694-699.
23. van Loon LJC. Application of protein or protein hydrolysates to improve postexercise recovery. Int J Sports Nutr Exerc Metab (2007) 17: S104-S117.
24.Calbet JA and Holst JJ. Gastric emptying, gastric secretion and enterogastrone response after administration of milk proteins or their peptide hydrosylates in humans. Eur J Nutr. (2004) 43:127-39.
25. Calbet JA et. al. Plasma glucagon and insulin responses depend on the rate of appearance of amino acids after ingestion of different protein solutions in humans. J Nutr. (2002) 132:2174-82.
26. Gropper SS et. al. Plasma amino acid response to ingestion of L-amino acids and whole protein. J Pediatr Gastroenterol Nutr. (1993) 16(2):143-50.
27. Boirie, Y. et. al. Slow and fast dietary proteins differently modulate postprandial protein accretion. Proc Natl Acad Sci USA (1997) 94: 14930-14935.28
28. Gema Fruhbeck .Slow and fast dietary proteins. Nature (1998) 391: 843-844.
29. Phillips, SM. Eating the Right Proteins to Support Muscle Mass with Exercise. Presentation at the 4th International Whey Conference, 2005.
30. Dangin M et. al. Influence of the protein digestion rate on protein turnover in young and elderly subjects. J Nutr. (2002) 132:3228S-33S.
31. Dangin M. et. al. Influence of protein digestion rate on protein turnover in young and elderly subjects. J Nutr (2002) 132: 3228S-3233S.
32. Dangin M et. al. The digestion rate of protein is an independent regulating factor of postprandial protein retention. Am J Physiol Endocrinol Metab. (2001) 280:E340-8.
33. Bos C et. al. Postprandial kinetics of dietary amino acids are the main determinants of their metabolism after soy or milk protein ingestion in humans. J Nutr (2003) 133: 1308-1315.
34. Fouillet H et. al. Peripheral and splanchnic metabolism of dietary nitrogen are differently affected by the protein source in humans as assessed by compartmental modeling. J Nutr (2002) 132: 125-133.
35. Tipton KD et. al. Ingestion of casein and whey proteins result in muscle anabolism after resistance exercise. Med Sci Sports Exerc. (2004) 36:2073-81.
36. Elliot TA et. al. Milk ingestion stimulates net muscle protein synthesis following resistance exercise. Med Sci Sports Exerc. (2006) 38:667-74.
37. Hartman JW et. al. Consumption of fat-free fluid milk after resistance exercise promtes greater lean mass accretion than does consumption of soy or carbohydrate in young, novice, make weighlifters. Am J Clin Nutr (2007) 86: 373-381.
38. Bilsborough S and N. Mann. A review of issues of dietary protein intake in humans. Int J Sports Nutr Exerc Metab (2006) 16: 129-152.
39. Rennie MJ. Control of muscle protein synthesis as a result of contractile activity and amino acid availability: implications for protein requirements. Int J Sport Nutr Exerc Metab. (2001) 11:S170-6.
40. Ferraris RP and JM Diamond. Specfic regulation of intestinal nutrient transporters by their dietary substrates. Ann Rev Physiol (1989) 51: 125-141.
41. Harris A. et. al. Rapid orocecal transit in chronically active persons with high energy intake. J Appl Physiol (1991) 70: 1550-1553.

42. Cunningham KM et. al. Gastrointestinal adaptation to diets of differing fat composition in human volunteers. Gut (1991) 32: 483-386.
43. Cunningham KM et. al. The effect of short-term dietary supplementation with glucose on gastric emptying on humans. Br J Nutr (1991) 65: 15-19.
44. Shi G et. al. Specific adaptation of gastric emptying to diets with differing protein content in the rat: is endogenous cholecystokinin implicated? Gut. (1997) 41(5):612-8.
45. Cuthbertson D et. al. Anabolic signaling deficits underlie amino acid resistance of wasting, aging muscle. FASEB J. (2005) 19(3):422-4.
46. Phillips SM et. al. A critical examination of dietary protein requirements, benefits, and excesses in athletes. Int J Sports Nutr Exerc Metab (2007) 17: S58-S76.

References cited for Chapter 3: Basic Protein Metabolism

1. JC Waterlow. Protein turnover with special reference to man. Q J Exp Phys (1984) 69: 409-438.
2. Welle S and Nair KS. Relationship of resting metabolic rate to body composition and protein turnover. Am J Physiol. (1990) 258(6 Pt 1):E990-8.
3. Young, V. 1987 McCollum lecture. Kinetics of human amino acid metabolism: nutritional implications and some lessons. Am J Clin Nutr (1987) 46: 709-725.
4. Waterlow, JC. Metabolic adaptations to low intakes of energy and protein. Ann Rev Nutr (1986) 6: 495-526.
5. Millward, DJ and Rivers, JPW. The need for indispensable amino acids: the concept of the anabolic drive. Diabetes/Metabolism Rev (1989) 5:191-211.
6. McNurlan, MA and Garlick, PJ. Influence of nutrient intake on protein turnover. Diabetes/Metabolism Rev (1989) 5: 165-189.
7. Tessari P et. al. Differential effects of hyperinsulinemia and hyperaminoacidemia on leucine-carbon metabolism in vivo. Evidence for distinct mechanisms in regulation of net amino acid deposition.J Clin Invest. (1987) 79:1062-9.
8. Rennie MJ et. al. Latency, duration and dose response relationships of amino acid effects on human muscle protein synthesis. J Nutr. (2002) 132(10):3225S-7S.
9. Biolo G et. al. Physiologic hyperinsulinemia stimulates protein synthesis and enhances transport of selected amino acids in human skeletal muscle. J Clin Invest. (1995) 95(2):811-9.
10. Bell JA et. al. Short-term insulin and nutritional energy provision do not stimulate muscle protein synthesis if blood amino acid availability decreases. Am J Physiol Endocrinol Metab. (2005) 289(6):E999-1006.
11. Rennie MJ et. al. Branched-chain amino acids as fuels and anabolic signals in human muscle. J Nutr. (2006) 136(1 Suppl):264S-8S.
12. Kimball SR, New functions for amino acids: effects on gene transcription and translation. Am J Clin Nutr. (2006) 83(2):500S-507S.
13. May, ME and Buse, MG. Effects of branch-chain amino acids on protein turnover. Diabetes/Metabolism Rev (1989) 5: 227-245.
14. Kadowaki M. and T. Kanazawa. Amino acids as regulators of proteolysis. J Nutr (2003) 133: 2052S-2056S.
15. Price GM et. al. Nitrogen homeostasis in man: influence of protein intake on the amplitude of diurnal cycling of body nitrogen. Clin Sci (Lond). (1994) 86:91-102.
16. Wagenmakers, AJ. Protein and amino acid metabolism in human muscle. Skeletal Muscle Metabolism in Exercise and Diabetes. ed. Richter et. al. Plenum Press: New York, 1998.
17. Fern, EB et. al. Effects of exaggerated amino acid and protein supply in man. Experientia (1991) 47: 168-172.
18. Millward, DJ et. al. Physical activity, protein metabolism and protein requirements. Proc Nutr Soc (1994) 53: 223-240.
19. Phillips SM et. al. A critical examination of dietary protein requirements, benefits, and excesses in athletes. Int J Sports Nutr Exerc Metab (2007) 17: S58-S76.

20. Oddoye, EA and Margen, S. Nitrogen balance studies in humans: long-term effect of high nitrogen intake on nitrogen accretion. J Nutr (1979) 109: 363-377.
21. Phillips SM, et. al. Mixed muscle protein synthesis and breakdown after resistance exercise in humans.Am J Physiol. (1997) 273:E99-107.

References cited for Chapter 4: Protein Requirements

1. Carpenter KJ. "Protein and Energy: A Study of Changing Ideas in Nutrition" (1994) Cambridge University Press.
2. JC Waterlow. Protein turnover with special reference to man. Q J Exp Phys (1984) 69: 409-438.
3. JC Waterlow. Metabolic adaptation to low intakes of energy and protein. Ann Rev Nutr (1986) 6: 495-526.
4. "Advanced nutrition and human metabolism, 2nd ed." James L. Groff, Sareen S. Gropper, Sara M. Hunt. West Publishing, 1995.
5. DJ Millward. Macronutrient intakes as determinants of dietary protein and amino acid adequacy. J Nutr (2004) 134: 1588s-1598s.
6. Richardson, DP et. al. Quantitativ effect of an isoenergetic exchange of fat for carbohydrate on dietary protein utilization in healthy young men. Am J Clin Nutr (1979) 32: 2217-2226.
7. McCargar LJ et. al. Dietary carbohydrate-to-fat ratio: influence on whole-body nitrogen retention, substrate utilization, and hormone response in healthy male subjects. Am J Clin Nutr. (1989) 49:1169-78.
8. National Research Council "Recommended dietary allowances 10th ed" National Academy Press, 1989.
9. Vernon Young and P.L. Pellet. Protein evaluation, amino acid scoring and the food and drug administration's proposed food labeling regulations. J Nutr (1990) 121: 145-150.
10. Millward DJ. Can we define indispensable amino acid requirements and assess protein quality in adults? J Nutr (1994) 124 (8 Suppl): 1509S-1516S.
11. Lemon, PWR et. al. Protein requirements and muscle mass/strength changes during intensive training in novice bodybuilders. J App Phys (1992) 73: 767-775.
12. Tarnopolsky, MA et. al. Evaluation of protein requirements for trained strength athletes. J Appl Physiol (1992) 73: 1986-1995.
13. Peter Lemon. Is increased dietary protein necessary or beneficial for individuals with a physically active lifestyle. Nutr Rev (1996) 54: S169-S175.
14. Lemon, PR and J.P. Mullin. Effect of initial muscle glycogen level on protein catabolism during exercise. J Appl Physiol (1980) 48: 624-629.
15. Phillips SM et. al. A critical examination of dietary protein requirements, benefits, and excesses in athletes. Int J Sports Nutr Exerc Metab (2007) 17: S58-S76.
16. Tarnopolsky M. Protein requirements for endurance athletes. Nutrition (2004) 20: 662-668.
17. Millward DJ. Optimal intakes of protein in the human diet. Proc Nutr Soc. (1999) 58:403-13.
18. Millward DJ. Protein and amino acid requirements of athletes. J Sports Sci. (2004) 22(1):143-4.
19. Kopple JD. Uses and limitations of the balance technique. JPEN J Parenter Enteral Nutr. (1987) 11(5 Suppl):79S-85S.
20. Butterfield GE. Whole-body protein utilization in humans. Med Sci Sports Exerc. (1987) 19(5 Suppl):S157-65.
21. RR Wolfe. Protein supplements and exercise. Am J Clin Nutr (2000) 72 (suppl): 551s-557s.
22. Millward DJ et. al. Physical activity, protein metabolism and protein requirements. Proc Nutr Soc. (1994) 53(1):223-40.
23. Phillips SM. Protein requirements and supplementation in strength sports. Nutrition (2004) 20: 689-695.

24. Phillips SM. Dietary protein for athletes: from requirements to metabolic advantage. Appl Physiol Nutr Metab. (2006) 31(6):647-54.
25. Tipton KD AND RR Wolfe. Protein and amino acids for athletes.J Sports Sci. (2004) 22(1):65-79.
26. Macdermid PW and SR Stannard. A whey-supplemented, high-protein diet versus a high-carbohydrate diet: effects on endurance cycling performance. Int J Sport Nutr Exerc Metab. (2006) 16(1):65-77.
27. Westerterp-Plantenga MS. The significance of protein in food intake and body weight regulation. Curr Opin Clin Nutr Metab Care. (2003) 6(6):635-8.
28. Layman DK. Protein quantity and quality at levels above the RDA improves adult weight loss. J Am Coll Nutr. (2004) 23(6 Suppl):631S-636S.
29. Johnston CS et. al. Postprandial thermogenesis is increased 100% on a high-protein, low-fat diet versus a high-carbohydrate, low-fat diet in healthy, young women. J Am Coll Nutr. (2002) 21(1):55-61.
30. Layman DK, Baum JI. Dietary protein impact on glycemic control during weight loss. J Nutr. (2004) 134(4):968S-73S.
31. Westerterp-Plantenga MS, et. al. High protein intake sustains weight maintenance after body weight loss in humans. Int J Obes Relat Metab Disord. (2004) 28(1):57-64.
32. Phillips SM et. al. Gender differences in leucine kinetics and nitrogen balance in endurance athletes. J Appl Physiol. (1993) 75(5):2134-41.
33. Cortright RN and TR Koves. Sex differences in substrate metabolism and energy homeostasis. Can J Appl Physiol. (2000) 25(4):288-311.
34. Hamadeh MJ, et. al. Estrogen supplementation reduces whole body leucine and carbohydrate oxidation and increases lipid oxidation in men during endurance exercise. J Clin Endocrinol Metab. (2005) 90(6):3592-9.
35. Lamont LS et. al. Menstrual cycle and exercise effects on protein catabolism. Med Sci Sports Exerc. (1987) 19(2):106-10.
36. Tiidus PM. Influence of estrogen on skeletal muscle damage, inflammation, and repair. Exerc Sport Sci Rev. (2003) 31(1):40-4.
37. Volek JS et. al. Nutritional Aspects of Women Strength Athletes. Br J Sports Med. (2006) 40(9):742-8
38. Miller BF et. al. No effect of menstrual cycle on myofibrillar and connective tissue protein synthesis in contracting skeletal muscle. Am J Physiol Endocrinol Metab. (2006) 290(1):E163-E168.

References cited for Chapter 5: Protein Quality

1. Hoffman JR and Falvo MJ. Protein – Which is best? J Sports Science Medicine (2004) 3: 118-130.
2. Young, VR. Soy protein in relation to human protein and amino acid nutrition. J Am Diet Assoc (1991) 91: 828-835.
3. Wagenmakers, AJM. Muscle amino acid metabolism at rest and during exercise: role in human physiology and metabolism. Exercise and Sports Science Reviews (1998) 26: 287-314.
4. Advanced Nutrition and Human Metabolism, 2nd ed. James L. Groff, Sareen S. Gropper, Sara M. Hunt. West Publishing Company, 1995.
5. Energy and protein requirements. Report of a joint FAO/WHO/UN expert consultation. WHO technical report series 724. Geneva: World Health Organizations, 1985.
6. Young, VR and El-Khoury, AE. Can amino acid requirements for nutritional maintenance in adult humans be approximated from the amino acid composition of body mixed proteins? Proc Natl Acad Sci USA (1995) 92: 300-304.
7. Millward, DJ. Metabolic demands for amion acids and the human dietary requirement: Millward and Rivers (1988) revisited. J Nutr (1998) 128: 2563S-2576S.

8.Pellett, PL and Young, VR. Nutritional evaluation of protein foods. United Nations University, 1980.
9. Poullain, MG et. al. Effect of whey proteins, their oligopeptide hydrosylates and free amino acid mixtures on growth and nitrogen retention in fed and starved rats. J Parenteral and Enteral Nutrition (1989) 13: 382-386.
10. Millward DJ Macronutrient intakes as determinants of dietary protein and amino acid adequacy. J Nutr. (2004) 134(6 Suppl):1588S-1596S.
11. Henley, EC. Food and Drug Administration's proposed labeling rules for protein. J Am Diet Assoc (1992) 92: 293-294, 296.
12. Young, VR. and Pellett, PL. Protein evaulation, amino acid scoring and the Food and Drug Administrations's proposed food labeling regulations. J Nutr (1991) 121:145-150.
13. Hernandez, M et. al. The protein efficiency ratio of 30:70 mixtures of animal:vegetables protein are similar or higher than those of the animal foods alone. J Nutr (1996) 126: 574-581.
14. Food and Agriculture Organization and World Health Organization (1990) protein quality evaluation. Report of a joint FAO/WHO expert consultation. Food and agriculture organization of the United Nations, Rome, Italy.
15. Young, VR and Pellett, PL. Plant proteins in relation to human protein and amino acid requirements. Am J Clin Nutr (1994) 59 (suppl): 1203S-1212S.

References cited for Chapter 6: Amino Acid Requirements

1. Bier, DM. Intrinsically difficult problems: The kinetics of body proteins and amino acids in man. Diabetes/Metabolism Rev (1989) 5:111-132.
2. Millward, J. Can we define indispensable AA requirements and assess protein quality in adults? J Nutr (1994) 1509s-1516s.
3. Smith GI et. al. Measurement of human mixed muscle protein fractional synthesis rate depends on the choice of amino acid tracer. Am J Physiol Endocrinol Metab. (2007) 293(3):E666-71.
4. Young, VR and El-Khoury, AE. Can amino acid requirements for nutritional maintenance in adult humans be approximate from the amino acid composition of mixed body proteins? Proc Natl Acad Sci, USA (1995) 92: 300-304.
5. FAO/WHO/UNU. Energy and protein requirements. Report of a joint FAO/WHO/UNU expert consultation. WHO Tech Report Ser 1985; 724.
6. Marchini JS et. al. Requirements for indispensable amino acids in adult humans: longer-term amino acid kinetic study with support for the adequacy of the Massachusetts Institute of Technology amino acid requirement protein. Am J Clin Nutr (1993) 58: 670-683.
7. Meguid, MM et. al. Leucine kinetics at graded leucine intakes in young men. Am J Clin Nutr (1986) 43: 770-780.
8. Meguid, MM et. al. Valine kinetics at graded valine kinetics in young men. Am J Clin Nutr (1986) 43: 781-786.
9. Meredith, CN et. al. Lysine kinetics at graded lysine intakes in young men. Am J Clin Nutr (1986) 43: 787-794.
10. Zhao, Xi-he et. al. Threonine kinetics at graded threonine intakes in young men. Am J Clin Nutr (1986) 43: 795-802.
11. Tipton KD and RR Wolfe. Protein and amino acids for athletes.J Sports Sci. (2004) 22 (1):65-79.
12. Young, VR and Pellett, PL. Plant proteins in relation to human protein and amino acid requirements. Am J Clin Nutr (1994) 59 (suppl): 1203S-1212S.
13. National Research Council. Recommended Dietary Allowances, 10th ed. National Academy Press, 1989.
14. Boza, JJ et. al. Nutritional value and antigenicity of two milk protein hydrosylates in rats and guinea pigs. J Nutr (1994) 124: 1978-1986.
15. Young, VR. Soy protein in relation to human protein and amino acid nutrition. J Am Diet Assoc (1991) 91: 828-835.

16. Wagenmakers, AJ. Protein and amino acid metabolism in human muscle. Skeletal Muscle Metabolism in Exercise and Diabetes. ed. Richter et. al. Plenum Press: New York, 1998.
17. MacLean DA et. al. Stimulation of muscle ammonia production during exercise following branched-chain amino acid supplementation in humans. J Physiol. (1996) 493(Pt 3): 909–922.
18. Millward, DJ et. al. Physical activity, protein metabolism and protein requirements. Proc Nutr Soc (1994) 53: 223-240.
19. Lemon P. Is increased dietary protein necessary or beneficial for individuals with a physically active lifestyle? Nutrition Reviews (1996) 54: S169-S175.
20. Phillips SM et. al. A critical examination of dietary protein requirements, benefits, and excesses in athletes. Int J Sports Nutr Exerc Metab (2007) 17: S58-S76.
21. Wagenmakers, AJ et. al. Carbohydrate supplementation, glycogen depletion, and amino acid metabolism during exercise. Am J Physiol (1991) 260: E833-E890.
22. Koopman R et. al. Co-ingestion of protein and leucine stimulates muscle protein synthesis rates to the same extent in young and elderly lean men. Am J Clin Nutr. (2006) 84(3):623-32.
23. Walsh, NP et. al. The effects of high-intensity intermittent exercise on the plasma concentrations of glutamine and organic acids. Eur J Appl Physiol (1998) 77:434-8
24. Hiscock, N and MacKinnon LT. A comparison of plasma glutamine concentration in athletes from different sports. Med Sci Sports Exerc (1998) 30: 1693-1696.
25. Gleeson, M et. al. The effect of severe eccentric exercise-induced muscle damage on plasma elastase, glutamine and zinc concentrations. Eur J Appl Physiol (1998) 77:543-6
26. Ha E, Zemel MB. Functional properties of whey, whey components, and essential amino acids: mechanisms underlying health benefits for active people (review). J Nutr Biochem. (2003) 14(5):251-8. Review.
27. Candow DG et. al. Effect of whey and soy protein supplementation combined with resistance training in young adults. Int J Sports Nut Exerc Metab (2006) 16: 233-244.
28. Phillips SM, et. al. Dietary protein to support anabolism with resistance exercise in young men. J Am Coll Nutr. (2005) 24(2):134S-139S.
29. Hartman JW et. al. Consumption of fat-free fluid milk after resistance exercise promotes greater lean mass accretion than does consumption of soy or carbohydrate in young, novice, male weightlifters. Am J Clin Nutr. (2007) 86(2):373-81.
30. Tipton KD Ingestion of casein and whey proteins result in muscle anabolism after resistance exercise. Med Sci Sports Exerc. (2004) 36(12):2073-81.
31. Appicelli P et. al. Does type of dietary protein supplementation affect muscle strength/size gains in adult bodybuilders. Can J Appl Physiol (1995) 20 (suppl): 1P.
32. Phillips, SM. Eating the Right Proteins to Support Muscle Mass with Exercise. Presentation at the 4th International Whey Conference, 2005.
33. Layman DK, Baum JI. Dietary protein impact on glycemic control during weight loss. J Nutr. (2004) 134(4):968S-73S.
34. Mourier, A et. al. Combined effect of caloric restriction and branched-chain amino acid supplementation on body composition and exercise performance in elite wrestlers. Int J Sports Med (1997) 18: 47-55.
35. Finn KJ et. al. Glutamine supplementation did not benefit athletes during short-term weight reduction. Journal of Sports Science and Medicine (2003) 2: 163-168.

References cited for Chapter 7: Meal Frequency

1. Jenkins DJ Carbohydrate tolerance and food frequency. Br J Nutr. (1997) 77 (Suppl 1):S71-81.
2. Mann J. Meal frequency and plasma lipids and lipoproteins. Br J Nutr. (1997) 77 (Suppl 1):S83-90.
3. Bellisle F et. al. Meal frequency and energy balance. Br J Nutr. (1997) 77 (Suppl 1):S57-70.

4. Speechly DP and R Buffenstein. Greater appetite control associated with an increased frequency of eating in lean males. Appetite. (1999) 33(3):285-97.
5. Speechly DP et. al. Acute appetite reduction associated with an increased frequency of eating in obese males. Int J Obes Relat Metab Disord. (1999) 23(11):1151-9.
6. Hawley JA et. al. Effect of meal frequency and timing on physical performance. Br J Nutr. (1997) 77 (Suppl 1):S91-103.
7. Capaldo, B et. al. Splanchnic and leg substrate exchange after ingestion of a natural mixed meal in humans. Diabetes (1999) 48: 958-966.
8. Layman DK. Protein quantity and quality at levels above the RDA improves adult weight loss. J Am Coll Nutr. (2004) 23(6 Suppl):631S-636S.
9. Irwin MI, Feeley RM. Frequency and size of meals and serum lipids, nitrogen and mineral retention, fat digestibility, and urinary thiamine and riboflavin in young women. Am J Clin Nutr. (1967) 20(8): 816-24.
10. Swindells YE, The metabolic response of young women to changes in the frequency of meals. Br J Nutr. (1968) 22(4): 667-80.
11. Young CM, Frequency of feeding, weight reduction, and body composition. J Am Diet Assoc. (1971) 59(5): 466-72.
12. Young CM Frequency of feeding, weight reduction, and nutrient utilization. J Am Diet Assoc. (1971) 59(5): 473-80.
13. Antoine JM, Rohr R, Gagey MJ, Bleyer RE, Debry G. Feeding frequency and nitrogen balance in weight-reducing obese women. Hum Nutr Clin Nutr. (1984) 38(1):31-8.
14. Garrow JS. The effect of meal frequency and protein concentration on the composition of the weight lost by obese subjects. Br J Nutr. (1981) 45(1):5-15.
15. Iwao S et. al. Effects of meal frequency on body composition during weight control in boxers. Scand J Med Sci Sports. (1996) 6(5):265-72.
16. Hansen Øyvind et. al. The effect of meal frequency on body composition during 12-weeks of strength training.
12th Annual Congress of the ECSS, 11–14 July 2007 (Abstract)
17. Bohe J et. al. Latency and duration of stimulation of human muscle protein synthesis during continuous infusion of amino acids. J Physiol. (2001) 532(Pt 2):575-9.
18. Rennie MJ et. al. Latency, duration and dose response relationships of amino acid effects on human muscle protein synthesis. J Nutr. (2002) 132(10):3225S-7S.
19. Cuthbertson D et. al. Anabolic signaling deficits underlie amino acid resistance of wasting, aging muscle. FASEB J. (2005) 19(3):422-4.
20. Phillips SM et.al. A critical examination of dietary protein requirements, benefits, and excesses in athletes. Int J Sports Nutr Exerc Metab (2007) 17: S58-S76.
21. Bohe J. et al. Human muscle protein synthesis is modulated by extracellular, not intramuscular amino acid availability: a dose-response study. J Physiol (2003) 552: 315-324.
22. Boirie, Y. et. al. Slow and fast dietary proteins differently modulate postprandial protein accretion. Proc Natl Acad Sci USA (1997) 94: 14930-14935.
23. Dangin M et. al. The digestion rate of protein is an independent regulating factor of postprandial protein retention. Am J Physiol Endocrinol Metab. (2001) 280: E340-8.
24. el-Khoury AE, The 24-h kinetics of leucine oxidation in healthy adults receiving a generous leucine intake via three discrete meals. Am J Clin Nutr. (1995) 62(3): 579-90.
25. Arnal MA et. al. Protein pulse feeding improves protein retention in elderly women. Am J Clin Nutr. (1999) 69:1202-8.
26. Arnal MA et. al. Protein feeding pattern does not affect protein retention in young women. J Nutr. (2000) 130:1700-4.
27. Jacob Wilson and Gabriel J. Wilson. Contemporary Issues in Protein Requirements and Consumption for Resistance Trained Athletes Journal of the International Society of Sports Nutrition. (2006) 3(1):7-27.
28. Mosoni L and Mirand PP. Type and timing of protein feeding to optimize anabolism. Curr Opin Clin Nutr Metab Care. (2003) 6(3):301-6.

29. Paddon-Jones D et. al. Exogenous amino acids stimulate human muscle anabolism without interfering with the response to mixed meal ingestion. Am J Physiol Endocrinol Metab. (2005) 288(4):E761-7.

References cited for Chapter 8: Nutrient Timing around Training

1. Millward DJ et. al. Physical activity, protein metabolism and protein requirements. Proc Nutr Soc. (1994) 53(1):223-40.
2. Tipton KD AND RR Wolfe. Protein and amino acids for athletes. J Sports Sci. (2004) 22(1):65-79.
3. Hargreaves M. et. al. Pre-exercise carbohydrate and fat ingestion: effects on metabolism and performance. J Sports Sci (2004) 22: 31-38.
4. Carli G, et. al. Changes in the exercise-induced hormone response to branched chain amino acid administration. Eur J Appl Physiol Occup Physiol. (1992) 64(3):272-7.
5. Brouns F et. al. Effect of carbohydrate intake during warming-up on the regulation of blood glucose during exercise. Int J Sports Med. (1989) 10 (Suppl 1):S68-75.
6. Haff GG et. al. Carbohydrate supplementation and resistance training. J Strength Cond Res. (2003) 17(1):187-96.
7. Haff G. Carbohydrate supplementation attenuates muscle glycogen loss during acute bouts of resistance exercise. Int J Sport Nutr Exerc Metab. (2000) 10(3):326-39.
8. Tipton KD et. al. Timing of amino acid-carbohydrate ingestion alters anabolic response of muscle to resistance exercise. Am J Physiol Endocrinol Metab. (2001) 281(2):E197-206.
9. Tipton KD et. al. Stimulation of net muscle protein synthesis by whey protein ingestion before and after exercise. Am J Physiol Endocrinol Metab. (2007) 292(1):E71-6.
10. Tipton K. Role of protein and hydrosylates before exercise. Int J Sports Nutr Exerc Metab (2007) 17: S77-S86.
11. Cribb PJ and Hayes A. Effects of supplement timing and resistance exercise on skeletal muscle hypertrophy. Med Sci Sports Exerc. (2006) 38(11):1918-25.
12. Jeukendrup AE. Carbohydrate intake during exercise and performance. Nutrition. (2004) 20(7-8):669-77.
13. Nieman DC. Nutrition, exercise, and immune system function. Clin Sports Med. (1999) 18(3):537-48.
14. McAnulty S et. al. Carbohydrate Effect: Hormone and Oxidative Changes. Int J Sports Med. (2007) May 11
15. Koopman R et. al. Combined ingestion of protein and carbohydrate improves protein balance during ultra-endurance exercise. Am J Physiol Endocrinol Metab. (2004) 287(4):E712-20.
16. Jeukendrup AE, Jentjens R. Oxidation of carbohydrate feedings during prolonged exercise: current thoughts, guidelines and directions for future research. Sports Med. (2000) 29(6):407-24.
17. Wallis GA et. al. Oxidation of combined ingestion of maltodextrins and fructose during exercise. Med Sci Sports Exerc. (2005) 37(3):426-32.
18. Shi X et. al. Effects of carbohydrate type and concentration and solution osmolality on water absorption. Med Sci Sports Exerc. (1995) 27(12):1607-15.
19. Haff GG et. al. Carbohydrate supplementation and resistance training. J Strength Cond Res. (2003) 17(1):187-96.
20. Bird SP et. al. Liquid carbohydrate/essential amino acid ingestion during a short-term bout of resistance exercise suppresses myofibrillar protein degradation. Metabolism. (2006) 55:570-7.
21. Baty JJ et. al. The effect of a carbohydrate and protein supplement on resistance exercise, performance, hormonal response, and muscle damage. J Str Cond Res (2007) 21: 321-329.
22. Saunders MJ. Coingestion of carbohydrate-protein during endurance exercise: influence on performance and recovery. Int J Sports Nutr Exerc Metab (2007) 17: S87-S103.

23. van Essen M and Gibala MJ. Failure of protein to improve time trial performance when added to a sports drink. Med Sci Sports Exerc. (2006) 38(8):1476-83.
24. Gibala MJ. Protein metabolism and endurance exercise. Sports Med. (2007) 37(4-5):337-40.
25. Romano-Ely BC et. al. Effect of an isocaloric carbohydrate-protein-antioxidant drink on cycling performance. Med Sci Sports Exerc. (2006) 38(9):1608-16.
26. Saunders MJ et. al. Consumption of an oral carbohydrate-protein gel improves cycling endurance and prevents postexercise muscle damage. J Str Cond Res (2007) 21: 678-684.
27. Hargreaves M. Muscle glycogen and metabolic regulation. Proc Nutr Soc. (2004) 63(2):217-20.
28. Burke LM etl al. Carbohydrates and fat for training and recovery. J Sports Sci. (2004) 22:15-30.
29. Jentjens R and A. Jeukendrup. Determinants of post-exercise glycogen synthesis during short-term recovery. Sports Med (2003) 33: 117-144.
30. Piehl Aulin K et. al. Muscle glycogen resynthesis rate in humans after supplementation of drinks containing carbohydrates with low and high molecular masses. Eur J Appl Physiol. (2000) 81(4):346-51.
31. Stephens FB et. al. Post-exercise ingestion of a unique, high molecular weight glucose polymer solution improves performance during a subsequent bout of cycling exercise Journal of Sports Sciences, (2007); 1 – 6.
32. Zawadski KM et. al. Carbohydrate-protein complex increases the rate of muscle glycogen storage after exercise. J Appl Physiol (1992) 72: 1854-1859.
33. van Loon LJ et. al. Ingestion of protein hydrolysate and amino acid-carbohydrate mixtures increases postexercise plasma insulin responses in men. J Nutr. (2000) 130(10):2508-13.
34. van Loon LJ et. al. Plasma insulin responses after ingestion of different amino acid or protein mixtures with carbohydrate. Am J Clin Nutr. (2000) 72(1):96-105.
35. Phillips SM, et. al. Mixed muscle protein synthesis and breakdown after resistance exercise in humans.Am J Physiol. (1997) 273:E99-107.
36. Koopman R et. al. Combined ingestion of protein and free leucine with carbohydrate increases postexercise muscle protein synthesis in vivo in male subjects. Am J Physiol (2005) 288:E645-53.
37. Bohe J Human muscle protein synthesis is modulated by extracellular, not intramuscular amino acid availability: a dose-response study. J Physiol. (2003) 552:315-24.
38. Borsheim E et. al., Essential amino acids and muscle protein recovery from resistance exercise. Am J Physiol Endocrinol Metab. (2002) 283(4):E648-57.
39. Tipton KD et. al. Postexercise net protein synthesis in human muscle from orally administered amino acids. Am J Physiol. (1999) 276:E628-34.
40. Chandler, R. M. et. al. Dietary supplements affect the anabolic hormones after weight-training exercise. J. Appl. Physiol. (1994) 76: 839-845.
41. Bird SP et. al. Independent and combined effects of liquid carbohydrate/essential amino acid ingestion on hormonal and muscular adaptations following resistance training in untrained men. Eur J Appl Physiol. (2006) 97(2):225-38.
42. Miller SL etl al. Independent and combined effects of amino acids and glucose after resistance exercise. Med Sci Sports Exerc. (2003) 35:449-55.
43. Chromiak JA et. al. Effect of a 10-week strength training program and recovery drink on body composition, muscular strength and endurance, and anaerobic power and capacity. Nutrition. (2004) 20(5):420-7.
44. Borsheim E et al. Effect of carbohydrate intake on net muscle protein synthesis during recovery from resistance exercise. J Appl Physiol. (2004) 96:674-8.
45. Borsheim E et. al. Effect of an amino acid, protein, and carbohydrate mixture on net muscle protein balance after resistance exercise. Int J Sport Nutr Exerc Metab. (2004) 14(3):255-71

46. Levenhagen DK et. al. Postexercise protein intake enhances whole-body and leg protein accretion in humans. Med Sci Sports Exerc. (2002) 34(5):828-37.
47. Luden ND et. a. Postexercise Carbohydrate-Protein-Antoxidant ingestion decreases plasma creatine kinase and muscle soreness. Int J Sports Nutr Exerc Metab (2007) 17: 109-123.
48. Manninen AH. Hyperinsulinaemia, hyperaminoacidaemia and post-exercise muscle anabolism: the search for the optimal recovery drink. Br J Sports Med. (2006) 40(11):900-5.
49. van Loon LJC. Application of protein or protein hydroslylates to improve post-exercise recovery. Int J Sports Nutr Exerc Metab (2007) 17: S104-S117.
50. Levenhagen DK et. al. Postexercise nutrient intake timing in humans is critical to recovery of leg glucose and protein homeostasis.Am J Physiol Endocrinol Metab. (2001) 280(6):E982-93.
51. Esmarck B et. al. Timing of postexercise protein intake is important for muscle hypertrophy with resistance training in elderly humans. J Physiol. (2001) 535(Pt 1):301-11.
52. Rasmussen BB et. al. An oral essential amino acid-carbohydrate supplement enhances muscle protein anabolism after resistance exercise. J Appl Physiol. (2000) 88(2):386-92.
53. Biolo G . et. al. Increased rates of muscle protein turnover and amino acid transport after resistance exercise in humans. Am J Physiol. (1995) 268(3 Pt 1):E514-20.
54. Tipton KD et. al. Ingestion of casein and whey proteins result in muscle anabolism after resistance exercise.Med Sci Sports Exerc. (2004) 36(12):2073-81.
55. Phillips, SM. Eating the Right Proteins to Support Muscle Mass with Exercise. Presentation at the 4th International Whey Conference, 2005.
56. Elliot TA et. al. Milk ingestion stimulates net muscle protein synthesis following resistance exercise. Med Sci Sports Exerc. (2006) 38(4):667-74.
57. Wilkinson SB et. al. Consumption of fluid skim milk promotes greater muscle protein accretion after resistance exercise than does consumption of an isonitrogenous and isoenergetic soy-protein beverage. Am J Clin Nutr. (2007) 85(4):1031-40.
58. Phillips SM et al. Dietary protein to support anabolism with resistance exercise in young men. J Am Coll Nutr. (2005) 24(2):134S-139S. Review.
59. Hartman JW et. al. Consumption of fat-free fluid milk after resistance exercise promtes greater lean mass accretion than does consumption of soy or carbohydrate in young, novice, make weighlifters. Am J Clin Nutr (2007) 86: 373-381.
60. Kerksick CM et. al. The effects of protein and amino acid supplementation on performance and training adaptations during ten weeks of resistance training. J Strength Cond Res. (2006) 20(3):643-53.
61. Ivy JL Early postexercise muscle glycogen recovery is enhanced with a carbohydrate-protein supplement. J Appl Physiol. (2002) 93(4):1337-44.
62. Karp JR et. al. Chocolate milk as a post-exercise recovery aid. Int J Sport Nutr Exerc Metab. (2006)16(1):78-91.
63. Shirreffs SM et al. Milk as an effective post-exercise rehydration drink. Br J Nutr (2007): Pg 1-8
64. Coleman E.Int J Sport Nutr. Update on carbohydrate: solid versus liquid. (1994) 4(2):80-8.
65. Bloomer RJ et. al. Effects of meal form and composition on plasma testosterone, cortisol, and insulin following resistance exercise. Int J Sport Nutr Exerc Metab. (2000) 10(4):415-24.
66. Tipton KD et. al. Acute response of net muscle protein balance reflects 24-h balance after exercise and amino acid ingestion. Am J Physiol Endocrinol Metab. (2003) 284:E76-89.
67. Andersen LL. The effect of resistance training combined with timed ingestion of protein on muscle fiber size and muscle strength. Metabolism. (2005) 54(2):151-6.

68. Koopman R et. al. Protein Ingestion Further Augments S6K1 Phosphorylation in Skeletal Muscle Following Resistance Type Exercise in Males. J Nutr. (2007) 137(8):1880-6.
69.Borghouts LB and Keizer HA. Exercise and insulin sensitivity: a review. Int J Sports Med. (2000) Jan;21(1):1-12.
70. Magkos F et. al. Improved insulin sensitivity after a single bout of exercise is curvilinearly related to exercise energy expenditure. Clin Sci (Lond). (2007). Jul 18; [Epub ahead of print]
71. Praet SF and van Loon LJ. Optimizing the therapeutic benefits of exercise in type 2 diabetes. J Appl Physiol. (2007) Jul 26; [Epub ahead of print]
72. Krzentowski G et. al. Clin Physiol. Metabolic adaptations in post-exercise recovery. (1982) 2(4):277-88.
73. Kimber NE, et. al. Skeletal muscle fat and carbohydrate metabolism during recovery from glycogen-depleting exercise in humans. J Physiol. (2003) 548(Pt 3):919-27.
74. Folch N Metabolic response to a large starch meal after rest and exercise: comparison between men and women. Eur J Clin Nutr. (2003) 57(9):1107-15.
75. Chandler RM et. al. Dietary supplements affect the anabolic hormones after weight-training exercise. J Appl Physiol. (1994) 76(2):839-45.
76. Kraemer WJ et. al. Hormonal responses to consecutive days of heavy-resistance exercise with or without nutritional supplementation. J Appl Physiol. (1998) 85(4):1544-55.
77. Kraemer WJ et. al. Androgenic responses to resistance exercise: effects of feeding and L-carnitine. Med Sci Sports Exerc. (2006) 38(7):1288-96.
78. McDonald L. A systematic examination of the pointlessness of long reference lists. Int J You Know You Don't Read These (2072) 987: 10024-11000.

<u>References cited for Chapter 9: Protein Controversies</u>

1. Martin WF et. al. Dietary protein intake and renal function. Nutr Metab (2005) 2: 25.
2. Millward DJ. Optimal intakes of protein in the human diet. Proc Nutr Soc. (1999) 58(2): 403-13.
3. Poortmans JR and Dellalieux O. Do regular high protein diets have potential health risks on kidney function in athletes? Int J Sport Nutr Exerc Metab. (2000) 10(1):28-38.
4. Dawson-Hughes B. Calcium and protein in bone health. Proc Nutr Soc. (2003) 62(2): 505-9.
5. Bonjour JP. Dietary protein: an essential nutrient for bone health. J Am Coll Nutr. (2005) 24(6 Suppl): 526S-36S.
6. Massey LK. Dietary animal and plant protein and human bone health: a whole foods approach. J Nutr. (2003) 133(3):862S-865S.
7. Dawson-Hughes B. Interaction of dietary calcium and protein in bone health in humans. J Nutr. (2003) 133(3):852S-854S.
8. Frassetto L et. al. Diet, evolution and aging--the pathophysiologic effects of the post-agricultural inversion of the potassium-to-sodium and base-to-chloride ratios in the human diet. Eur J Nutr. (2001) 40(5):200-13.
9. Wiederkehr M, Krapf R. Metabolic and endocrine effects of metabolic acidosis in humans. Swiss Med Wkly. (2001) 131(9-10):127-32.
10. Barzel US and LK Massey Excess dietary protein can adversely affect bone. J Nutr. (1998) 128(6):1051-3.
11. Fogelholm M. Dairy products, meat and sports performance. Sports Med. (2003) 33(8):615-31.
12. Sabate J. The contribution of vegetarian diets to human health. Forum Nutr. (2003) 56:218-20.
13. Li D et. al. Lean meat and heart health. Asia Pac J Clin Nutr. (2005) 14(2):113-9.
14. Hodgson JM et. al. Increased lean red meat intake does not elevate markers of oxidative stress and inflammation in humans. J Nutr. (2007) 137(2):363-7. Links
15. Biesalski HK.Meat and cancer: meat as a component of a healthy diet.

Eur J Clin Nutr. (2002) 56 Suppl 1:S2-11.
16. Hodgson JM et. al. Partial substitution of carbohydrate intake with protein intake from lean red meat lowers blood pressure in hypertensive persons.Am J Clin Nutr. (2006) 83(4):780-7.
17. Hill M. Meat, cancer and dietary advice to the public. Eur J Clin Nutr. (2002) 56 Suppl 1:S36-41
18. Elmadfa I, Freisling H. Fat intake, diet variety and health promotion. Forum Nutr. (2005) (57):1-10.

References cited for Chapter 10: Whole Food Protein

1. Uhe AM A comparison of the effects of beef, chicken and fish protein on satiety and amino acid profiles in lean male subjects. J Nutr. (1992) 122(3):467-72.
2. Levenson CW and D. Axelrad. Too much of a good thing? Update on fish consumption and mecury exposure. Nutrition Rev (2006) 64: 139-145.
3. Micheletti A et. al. Zinc status in athletes: relation to diet and exercise. Sports Med. (2001) 31(8):577-82.
4. Suedekum NA et. al. Iron and the athlete. Curr Sports Med Rep. (2005) 4(4):199-202.
5. Hurrell RF Meat protein fractions enhance nonheme iron absorption in humans. J Nutr. (2006) 136(11):2808-12.
6. Mann N. Dietary lean red meat and human evolution. Eur J Nutr. (2000) 39:71-9
7. Li D et. al. Lean meat and heart health. Asia Pac J Clin Nutr. (2005) 14:113-9.
8. Taubes G. Nutrition. The soft science of dietary fat. Science. (2001) 291(5513):2536-45.
9. Lambert CP et. al. Macronutrient considerations for the sport of bodybuilding. Sports Med. (2004) 34:317-27.
10. Fernandez ML. Dietary cholesterol provided by eggs and plasma lipoproteins in healthy populations. Curr Opin Clin Nutr Metab Care. (2006) 9(1):8-12. Review.
11. Kritchevsky SB. A review of scientific research and recommendations regarding eggs. J Am Coll Nutr. (2004) 23(6 Suppl):596S-600S.
12. Nuttall FQ, Gannon MC. Metabolic response to egg white and cottage cheese protein in normal subjects. Metabolism. (1990) 39:749-55.
13. Shah NP. Effects of milk-derived bioactives: An Overview. Br J Nutr (2000) 84 (Suppl 1): S3-S10.
14. Layman DK. The role of leucine in weight loss diets and glucose homeostasis. J Nutr. (2003) 133:261S-267S.
15. Gill HS and ML Cross. Anticancer properties of bovine milk. Br J Nutr (2000) 84 (Suppl 1): S161-S166.
16. Cross ML and HS Gill. Immunomodulatory properties of milk. Br J Nutr (2000) 84 (suppl 1): S81-S89.
17. Navarro V et. al. The body fat-lowering effect of conjugated linoleic acid: a comparison between animal and human studies. J Physiol Biochem. (2006) 62(2):137-47.
18. Miller GD et. al. The importance of meeting calcium needs with foods. J Am Coll Nutr (2001) 20: 168s-185s.
19. Zemel MB. Role of calcium and dairy products in energy partitioning and weight management. Am J Clin Nutr (2004). 79 (suppl): 907s-912s.
20. Bahna SL Cow's milk allergy versus cow milk intolerance.. Ann Allergy Asthma Immunol. (2002) 89(6 Suppl 1):56-60.
21. Crittenden RG and LE Bennett. Cow's milk allergy: a complex disorder. J Am Coll Nutr. (2005) 24(6 Suppl):582S-91S.
22. Huth PJ et. al. Major scientific advances with dairy foods in nutrition and health. J Dairy Sci. (2006) 89(4):1207-21.
23. Guarner F et. al. Should yoghurt cultures be considered probiotic? Br J Nutr. (2005) 93(6):783-6.

24. Elliot TA et. al. Milk ingestion stimulates net muscle protein synthesis following resistance exercise. Med Sci Sports Exerc. (2006) 38(4):667-74.
25. Wilkinson SB et. al. Consumption of fluid skim milk promotes greater muscle protein accretion after resistance exercise than does consumption of an isonitrogenous and isoenergetic soy-protein beverage. Am J Clin Nutr. (2007) 85(4):1031-40.
26. Karp JR et. al. Chocolate milk as a post-exercise recovery aid. Int J Sport Nutr Exerc Metab. (2006) 16(1):78-91.
27. Shirreffs SM et al. Milk as an effective post-exercise rehydration drink. Br J Nutr (2007): Pg 1-8
28. Messina, M. Legumes and soybeans: overview of their nutritional profiles and health effects. Am J Clin Nutr (1999) 70 (suppl): 439s-450s.
29.Garcia-Lorda P. Nut consumption, body weight and insulin resistance. Eur J Clin Nutr. (2003) 57 Suppl 1:S8-11.
30. Sabate J. Nut consumption and body weight. Am J Clin Nutr (2003) 78 (suppl): 647s-650s.
31. St-Onge M. Dietary fats, teas, dairy and nuts: potential functional foods for weight control? Am J Clin Nutr (2005) 81: 7-15.
32. Mindy S. Kurzer Hormonal Effects of Soy in Premenopausal Women and Men J. Nutr. (2002) 132:570S-573S.
33. Reinwald S. and CM Weaver. Soy isoflavones and bone health: a double-edged sword? J Nat Prod. (2006) 69:450-9.
34. Sirtori CR et. al. Phytoestrogens: end of a tale? Ann Med. (2005) 37:423-38.
35. Kurzer MS. Hormonal effects of soy in premenopausal women and men. J Nutr. (2002) 132(3):570S-573S.
36. Goodin S et. al. Clinical and biological activity of soy protein powder supplementation in healthy male volunteers. Cancer Epidemiology Biomarkers & Prevention (2007).16: 829-833.
37. Messina M and G. Redmond. Effects of soy protein and soybean isoflavones on thyroid function in healthy adults and hypothyroid patients: a review of the relevant literature. Thyroid. (2006) 16:249-58.
38. Forsythe WA 3rd. Soy protein, thyroid regulation and cholesterol metabolism. J Nutr. (1995) 125(3 Suppl):619S-623S.
39. Doerge DR and Sheehan DM. Goitrogenic and estrogenic activity of soy isoflavones. Environ Health Perspect. (2002) 110 (Suppl 3):349-53.
40. Ortolani C and Pastorello EA. Food allergies and food intolerances. Best Pract Res Clin Gastroenterol. (2006) 20(3):467-83.
41. Sicherer SH and Sampson HA. 9. Food allergy. J Allergy Clin Immunol. (2006) 117(2 Suppl Mini-Primer):S470-5.
42. Schuppan D et. al. Celiac disease: epidemiology, pathogenesis, diagnosis, and nutritional management. Nutr Clin Care. (2005) 8(2):54-69.

References cited for Chapter 11: Protein powders

1. Bucci LR and L. Unlu. Protein and amino acid supplements in exercise and sport. 191-212. In Energy-Yielding Macronutrients and Exercise Metabolism in Sports Nutrition, edited by Driskell JA. and Wolinsky I.
2. Norton LE, Layman DK. Leucine regulates translation initiation of protein synthesis in skeletal muscle after exercise. J Nutr. (2006)136(2):533S-537S.
3. Layman DK. The role of leucine in weight loss diets and glucose homeostasis. J Nutr. (2003) 133(1):261S-267S.
4. Boirie Y et. al. Slow and fast dietary proteins differently modulate postprandial protein accretion. Proc Natl Acad Sci U S A. (1997) 94(26):14930-5.
5. DeSanti L Effect of a hypocaloric diet, increased protein intake and resistance training on lean mass gains and fat mass loss in overweight police officers. Ann Nutr Metab. (2000) 44(1):21-9.

6. Cribb PJ The effect of whey isolate and resistance training on strength, body composition, and plasma glutamine. Int J Sport Nutr Exerc Metab. (2006) 16(5):494-509.
7. Hall WL et. al. Casein and whey exert different effects on plasma amino acid profiles, gastrointestinal hormone secretion and appetite. Br J Nutr. (2003) 89(2):239-48
8.Lands LC Effect of supplementation with a cysteine donor on muscular performance. J Appl Physiol. (1999) 87(4):1381-5.
9. Middleton N et. al. Whole blood and mononuclear cell glutathione response to dietary whey protein supplementation in sedentary and trained male human subjects. Int J Food Sci Nutr. (2004) 55(2):131-41.
10. Kinscherf R et. al. Low plasma glutamine in combination with high glutamate levels indicate risk for loss of body cell mass in healthy individuals: the effect of N-acetyl-cysteine. J Mol Med. (1996) 74(7):393-400.
11. Marshall K. Therapeutic applications of whey protein. Altern Med Rev. (2004) 9(2):136-56.
12. Huth PJ et. al. Major scientific advances with dairy foods in nutrition and health. J Dairy Sci. (2006) 89(4):1207-21.
13. Phillips, SM. Eating the Right Proteins to Support Muscle Mass with Exercise. Presentation at the 4th International Whey Conference, 2005.
14. Wilkinson SB et. al. Consumption of fluid skim milk promotes greater muscle protein accretion after resistance exercise than does consumption of an isonitrogenous and isoenergetic soy-protein beverage. Am J Clin Nutr. (2007) 85(4):1031-40.
15. Lacroix M et. al. Compared with casein or total milk protein, digestion of milk soluble proteins is too rapid to sustain the anabolic postprandial amino acid requirement. Am J Clin Nutr. (2006) 84(5):1070-9.
16. Kerksick CM et. al. The effects of protein and amino acid supplementation on performance and training adaptations during ten weeks of resistance training. J Strength Cond Res. (2006) 20(3):643-53.
17. Elliot TA et. al. Milk ingestion stimulates net muscle protein synthesis following resistance exercise. Med Sci Sports Exerc. (2006) 38(4):667-74.
18.Hill S et. al. Moderate intensity resistance exercise, plus or minus soy intake: effects on serum lipid peroxides in young adult males. Int J Sport Nutr Exerc Metab. (2004) 14(2):125-32.
19. Box W. Soy intake plus moderate weight resistance exercise: effects on serum concentrations of lipid peroxides in young adult women. J Sports Med Phys Fitness. (2005) 45(4):524-8.
20. Candow DG et. al. Effect of whey and soy protein supplementation combined with resistance training in young adults. Int J Sport Nutr Exerc Metab. (2006) 16(3):233-44.
21. Phillips SM. Dietary protein to support anabolism with resistance exercise in young men. J Am Coll Nutr. (2005) 24(2):134S-139S.
22. Hartman JW et. al. Consumption of fat-free fluid milk after resistance exercise promotes greater lean mass accretion than does consumption of soy or carbohydrate in young, novice, male weightlifters. Am J Clin Nutr. 2007 Aug;86(2):373-81.
23. Luiking YC et. al. Casein and soy protein meals differentially affect whole-body and splanchnic protein metabolism in healthy humans. J Nutr. (2005) 135(5):1080-7.
24. Bos C et. al. Postprandial kinetics of dietary amino acids are the main determinant of their metabolism after soy or milk protein ingestion in humans. J Nutr. (2003) 133(5):1308-15.

References cited for Chapter 12: Amino acid supplements

1. Mittendorfer, et. al. Whole body and skeletal muscle glutamine metabolism in healthy subjects. Am J Physiol Endocrinol Metab (2001) 280: E323–E333,
2. MacLennan PA et. al. A positive relationship between protein synthetic rate and intracellular glutamine concentration in perfused rat skeletal muscle. FEBS Lett. (1987) 215(1):187-91.

3. MacLennan PA et. al. Inhibition of protein breakdown by glutamine in perfused rat skeletal muscle. FEBS Lett. (1988) 237(1-2):133-6.
4. Zachwieja JJ, et. al. Intravenous glutamine does not stimulate mixed muscle protein synthesis in healthy young men and women. Metabolism. (2000) 49(12):1555-60.
5. Svanberg E et. al. The effect of glutamine on protein balance and amino acid flux across arm and leg tissues in healthy volunteers. Clin Physiol. (2001) 21(4):478-89.
6. Candow DG Effect of glutamine supplementation combined with resistance training in young adults. Eur J Appl Physiol. (2001) 86(2):142-9.
7. Antonio J et. al. The effects of high-dose glutamine ingestion on weightlifting performance. J Strength Cond Res. (2002) 16(1):157-60.
8. Wilkinson SB et. al. Addition of glutamine to essential amino acids and carbohydrate does not enhance anabolism in young human males following exercise. Appl Physiol Nutr Metab. (2006) 31(5):518-29.
9. Iwashita et. al. Glutamine supplementation increases postprandial energy expenditure and fat oxidation in humans. JPEN J Parenter Enteral Nutr. (2006) 30(2):76-80.
10. Finn KJ et. al. Glutamine supplementation did not benefit athletes during short-term weight reduction. Journal of Sports Science and Medicine (2003) 2: 163-168.
11. Melis GC et. al. Glutamine: recent developments in research on the clinical significance of glutamine. Curr Opin Clin Nutr Metab Care. (2004) 7(1):59-70.
12. Castell L. Glutamine supplementation in vitro and in vivo, in exercise and in immunodepression.Sports Med. (2003) 33(5):323-45.
13. Castell LM Does glutamine have a role in reducing infections in athletes? Eur J Appl Physiol Occup Physiol. (1996) 73(5):488-90.
14. Walsh NP Effect of oral glutamine supplementation on human neutrophil lipopolysaccharide-stimulated degranulation following prolonged exercise. Int J Sport Nutr Exerc Metab. (2000) 10(1):39-50.
15. Rohde T Competitive sustained exercise in humans, lymphokine activated killer cell activity, and glutamine--an intervention study. Eur J Appl Physiol Occup Physiol. (1998) 78(5):448-53.
16. Petibois C et. al. Biochemical aspects of overtraining in endurance sports: a review. Sports Med. (2002) 32(13):867-78.
17. Walsh NP, et. al. Glutamine, exercise and immune function. Links and possible mechanisms. Sports Med. (1998) 26(3):177-91.
18. Holecek M. Relation between glutamine, branched-chain amino acids, and protein metabolism. Nutrition. (2002) 18(2):130-3.
19. Bacurau RF et. al. Carbohydrate supplementation during intense exercise and the immune response of cyclists. 1: Clin Nutr. (2002) 21(5):423-9.
20. van Hall Get. al. The effect of free glutamine and peptide ingestion on the rate of muscle glycogen resynthesis in man. Int J Sports Med. (2000) 21(1):25-30.
21. Fernstrom JD. Branched-chain amino acids and brain function. J Nutr. (2005) 135(6 Suppl):1539S-46S.
22. Sutton EE et. al. Ingestion of tyrosine: effects on endurance, muscle strength, and anaerobic performance. Int J Sport Nutr Exerc Metab. (2005) 15(2):173-85.
23. Chinevere TD et. al. Effects of L-tyrosine and carbohydrate ingestion on endurance exercise performance. J Appl Physiol. (2002) 93(5):1590-7.
24. Hull KM and TJ Maher. L-tyrosine potentiates the anorexia induced by mixed-acting sympathomimetic drugs in hyperphagic rats. J Pharmacol Exp Ther. (1990) 255(2):403-9.
25. Fernstrom JD. Branched-chain amino acids and brain function. J Nutr. (2005) 135(6 Suppl):1539S-46S.
26. Layman DK, Potential importance of leucine in treatment of obesity and the metabolic syndrome. J Nutr. (2006) 136(1 Suppl):319S-23S.
27. Mourier A Combined effects of caloric restriction and branched-chain amino acid supplementation on body composition and exercise performance in elite wrestlers. Int J Sports Med. (1997) Jan;18(1):47-55.

28. Calder PC. Branched-chain amino acids and immunity. J Nutr. (2006) 136(1 Suppl):288S-93S.
29. Bassit RA et. al. Branched-chain amino acid supplementation and the immune response of long-distance athletes. Nutrition. (2002) 18(5):376-9.
30. Blomstrand E. A role for branched-chain amino acids in reducing central fatigue J Nutr. (2006) 136(2):544S-547S.
31. Meeusen R and Watson P. Amino acids and the brain: do they play a role in "central fatigue"? Int J Sports Nutrition Exerc Metab (2007) 17: S37-S46.
32. Meeusen R et. al. Central fatigue: the serotonin hypothesis and beyond. Sports Med. (2006) 36(10):881-909.
33. MacLean DA et. al. Stimulation of muscle ammonia production during exercise following branched-chain amino acid supplementation in humans. J Physiol. (1996) 493(Pt 3): 909–922.
34. Colombani PC et. al. Metabolic effects of a protein-supplemented carbohydrate drink in marathon runners. Int J Sport Nutr. (1999) 9(2):181-201.
35. Mero A. Leucine supplementation and intensive training. Sports Med. (1999) 27(6):347-58.
36. Carli G, et. al. Changes in the exercise-induced hormone response to branched chain amino acid administration. Eur J Appl Physiol Occup Physiol. (1992); 64(3):272-7.
37. Coombes JS et. al. Effects of branched-chain amino acid supplementation on serum creatine kinase and lactate dehydrogenase after prolonged exercise. J Sports Med Phys Fitness.(2000) 40(3):240-6.
38. Tang FC. Influence of branched-chain amino acid supplementation on urinary protein metabolite concentrations after swimming. J Am Coll Nutr. (2006) 25(3):188-94.
39. Crowe MJ et. al. Effects of dietary leucine supplementation on exercise performance. Eur J Appl Physiol. (2006) 97(6):664-72.
40. Blomstrand E Branched-chain amino acids activate key enzymes in protein synthesis after physical exercise. J Nutr. (2006) 136(1 Suppl):269S-73S.
41. Kimball SR, New functions for amino acids: effects on gene transcription and translation. Am J Clin Nutr. (2006) 83(2):500S-507S.
42. Karlsson HK et. al. Branched-chain amino acids increase p70S6k phosphorylation in human skeletal muscle after resistance exercise. Am J Physiol Endocrinol Metab. (2004) 287(1):E1-7.
42. Koopman R et. al. Combined ingestion of protein and free leucine with carbohydrate increases postexercise muscle protein synthesis in vivo in male subjects. Am J Physiol Endocrinol Metab. (2005) 288(4):E645-53.
44. Mero A et. al. Leucine supplementation and serum amino acids, testosterone, cortisol and growth hormone in male power athletes during training. J Sports Med Phys Fitness. (1997) 37(2):137-45.
45. Pitkanen HT et. al. Leucine supplementation does not enhance acute strength or running performance but affects serum amino acid concentration. Amino Acids. (2003) 25(1):85-94
46. Pitkanen H et. al. Serum amino acid responses to three different exercise sessions in male power athletes. J Sports Med Phys Fitness. (2002) 42(4):472-80.
47. 29a. The abstract for the Italian BCAA paper can be found at the following online adress:
http://www.nutritionexpress.com/article+index/authors/edmund+r+burke+phd/showarticle.aspx?articleid=7
48. Nissen S et. al. Effect of leucine metabolite beta-hydroxy-beta-methylbutyrate on muscle metabolism during resistance-exercise training. J Appl Physiol. (1996) 81(5):2095-104.
49. van Someren KA et. al. Supplementation with beta-hydroxy-beta-methylbutyrate (HMB) and alpha-ketoisocaproic acid (KIC) reduces signs and symptoms of exercise-induced muscle damage in man. Int J Sport Nutr Exerc Metab. (2005) 15(4):413-24.

50. Kreider RB et. al. Effects of calcium beta-hydroxy-beta-methylbutyrate (HMB) supplementation during resistance-training on markers of catabolism, body composition and strength. Int J Sports Med. (1999) 20(8):503-9.
51. Hoffman JR et. al. Effects of beta-hydroxy beta-methylbutyrate on power performance and indices of muscle damage and stress during high-intensity training. J Strength Cond Res. (2004) 18(4):747-52.
52. Knitter AE et. al. Effects of beta-hydroxy-beta-methylbutyrate on muscle damage after a prolonged run. J Appl Physiol. (2000) 89(4):1340-4.
53. Lamboley CR et. al. Effects of beta-hydroxy-beta-methylbutyrate on aerobic-performance components and body composition in college students. Int J Sport Nutr Exerc Metab. (2007) 17(1):56-69.
54. O'Connor DM, et. al.Effects of beta-hydroxy-beta-methylbutyrate and creatine monohydrate supplementation on the aerobic and anaerobic capacity of highly trained athletes. J Sports Med Phys Fitness. (2003) 43(1):64-8.
55. Tipton KD et. al. Nonessential amino acids are not necessary to stimulate net muscle protein synthesis in healthy volunteers. J Nutr Biochem. (1999) 10(2):89-95.
56. Cuthbertson D et. al. Anabolic signaling deficits underlie amino acid resistance of wasting, aging muscle. FASEB J. (2005) 19(3):422-4.
57. Paddon-Jones D et. al. Exogenous amino acids stimulate human muscle anabolism without interfering with the response to mixed meal ingestion. Am J Physiol Endocrinol Metab. (2005) 288(4):E761-7.
58. Borsheim E et. al. Essential amino acids and muscle protein recovery from resistance exercise. Am J Physiol Endocrinol Metab. (2002) 283(4):E648-57.
59. Tipton KD et. al. Postexercise net protein synthesis in human muscle from orally administered amino acids. Am J Physiol. (1999) 276(4 Pt 1):E628-34.
60. Rasmussen BB et. al. An oral essential amino acid-carbohydrate supplement enhances muscle protein anabolism after resistance exercise. J Appl Physiol. (2000) 88(2):386-92.
61. Bird SP et. al. Liquid carbohydrate/essential amino acid ingestion during a short-term bout of resistance exercise suppresses myofibrillar protein degradation. Metabolism. (2006) 55(5):570-7.
62. Tipton KD Timing of amino acid-carbohydrate ingestion alters anabolic response of muscle to resistance exercise. Am J Physiol Endocrinol Metab. (2001) Aug;281(2):E197-206.
63. Tipton KD et. al. Stimulation of net muscle protein synthesis by whey protein ingestion before and after exercise. Am J Physiol Endocrinol Metab. (2007) 292(1):E71-6.
64. Antonio J. Effects of exercise training and amino-acid supplementation on body composition and physical performance in untrained women. Nutrition. (2000) 16(11-12):1043-6.
65. Bemben MG, Lamont HS Creatine supplementation and exercise performance: recent findings. Sports Med. (2005) 35(2):107-25.
66. van Loon LJ et. al. Effects of creatine loading and prolonged creatine supplementation on body composition, fuel selection, sprint and endurance performance in humans. Clin Sci (Lond). (2003) 104(2):153-62.
67. Rawson ES, Volek JS. Effects of creatine supplementation and resistance training on muscle strength and weightlifting performance. J Strength Cond Res. (2003) 17(4):822-31.
68. Kreider RB. Effects of creatine supplementation on performance and training adaptations. Mol Cell Biochem. (2003) 244(1-2):89-94.
69. Willoughby DS and J. Rosene. Effects of oral creatine and resistance training on myosin heavy chain expression. Med Sci Sports Exerc. (2001) 33(10):1674-81.
70. Willoughby DS and JM Rosene. Effects of oral creatine and resistance training on myogenic regulatory factor expression. Med Sci Sports Exerc. (2003) 35(6):923-9.
71. Parise G et. al. Effects of acute creatine monohydrate supplementation on leucine kinetics and mixed-muscle protein synthesis. J Appl Physiol. (2001) 91(3):1041-7.

72. Mihic S et. al. Acute creatine loading increases fat-free mass, but does not affect blood pressure, plasma creatinine, or CK activity in men and women. Med Sci Sports Exerc. (2000) 32(2):291-6.
73. Branch JD. Effect of creatine supplementation on body composition and performance: a meta-analysis. Int J Sport Nutr Exerc Metab. (2003) 13(2):198-226.
74. Louis M et. al. No effect of creatine supplementation on human myofibrillar and sarcoplasmic protein synthesis after resistance exercise. Am J Physiol Endocrinol Metab. (2003) 285(5):E1089-94.
75. Louis M et. al. Creatine supplementation has no effect on human muscle protein turnover at rest in the postabsorptive or fed states. Am J Physiol Endocrinol Metab. (2003) 284(4):E764-70.
76. Juhn MS, Oral creatine supplementation and athletic performance: a critical review. Clin J Sport Med. (1998) 8(4):286-97.
77. Chwalbinska-Moneta J. Effect of creatine supplementation on aerobic performance and anaerobic capacity in elite rowers in the course of endurance training. Int J Sport Nutr Exerc Metab. (2003) 13(2):173-83.
78. Engelhardt M et. al. Creatine supplementation in endurance sports. Med Sci Sports Exerc. (1998) 30(7):1123-9.
79. Santos RV et. al. The effect of creatine supplementation upon inflammatory and muscle soreness markers after a 30km race. Life Sci. (2004) 75(16):1917-24.
80. van Loon LJ et. al. Creatine supplementation increases glycogen storage but not GLUT-4 expression in human skeletal muscle. Clin Sci (Lond). (2004) 106(1):99-106.
81. Syrotuik DG, Bell GJ.Acute creatine monohydrate supplementation: a descriptive physiological profile of responders vs. nonresponders. J Strength Cond Res. (2004) 18(3):610-7.
82. Hespel P et. al. Opposite actions of caffeine and creatine on muscle relaxation time in humans. J Appl Physiol. (2002) 92(2):513-8.
83. Vandenberghe K, et. al.Caffeine counteracts the ergogenic action of muscle creatine loading. J Appl Physiol. (1996) 80(2):452-7.
84. van Leemputte M et. al. Shortening of muscle relaxation time after creatine loading. J Appl Physiol. (1999) 86(3):840-4.
85. Bizzarini E et. al. Is the use of oral creatine supplementation safe? J Sports Med Phys Fitness. (2004) 44(4):411-6.
86. Greenwood M, et. al. Creatine supplementation during college football training does not increase the incidence of cramping or injury. Mol Cell Biochem. (2003) 244(1-2):83-8.
87. Casey A and PL Greenhaff PL. Does dietary creatine supplementation play a role in skeletal muscle metabolism and performance? Am J Clin Nutr. (2000) 72(2 Suppl):607S-17S.
88. Steenge GR et. al. Protein- and carbohydrate-induced augmentation of whole body creatine retention in humans. J Appl Physiol. (2000) 89(3):1165-71.
89. Burke DG et. al. Effect of alpha-lipoic acid combined with creatine monohydrate on human skeletal muscle creatine and phosphagen concentration. Int J Sport Nutr Exerc Metab. (2003) 13(3):294-302.
90. Preen D et. al. Int J Sport Nutr Exerc Metab. Creatine supplementation: a comparison of loading and maintenance protocols on creatine uptake by human skeletal muscle. (2003) 13(1):97-111.
91. Burke DG et. al. Effect of creatine and weight training on muscle creatine and performance in vegetarians. Med Sci Sports Exerc. (2003) 35(11):1946-55.
92. Tarnopolsky, M. et. al. Acute and moderate-term creatine monohydrate supplementation does not affect creatine transporter mRNA or protein content in either young or elderly humans. Mol Cell Biochem. (2003) 244(1-2):159-66.
93. Karlic H, Lohninger A. Supplementation of carnitine in athletes: does it make sense? Nutrition. (2004) 20(7-8):709-15.

94. Brass EP. Carnitine and sports medicine: use or abuse? Ann N Y Acad Sci. (2004) 1033:67-78.
95. Stephens FB. New insights concerning the role of carnitine in the regulation of fuel metabolism in skeletal muscle. J Physiol. (2007) Mar 1; [Epub ahead of print]
96. Brass EP. Supplemental carnitine and exercise. Am J Clin Nutr. (2000) 72(2 Suppl):618S-23S.
97. Stephens FB et. al. Carbohydrate ingestion augments L-carnitine retention in humans. J Appl Physiol. (2007) 102(3):1065-70.
98. Stephens FB et. al. A threshold exists for the stimulatory effect of insulin on plasma L-carnitine clearance in humans. Am J Physiol Endocrinol Metab. (2007) 292(2):E637-41.
99. Stephens FB et. al. An acute increase in skeletal muscle carnitine content alters fuel metabolism in resting human skeletal muscle. J Clin Endocrinol Metab. (2006) 91(12):5013-8.
100. Volek JS et. al. L-Carnitine L-tartrate supplementation favorably affects markers of recovery from exercise stress. Am J Physiol Endocrinol Metab. (2002) 282(2):E474-82.
101. Kraemer WJ et. al. The effects of L-carnitine L-tartrate supplementation on hormonal responses to resistance exercise and recovery. J Strength Cond Res. (2003) 17(3):455-62.
102. Spiering BA et. al. Responses of criterion variables to different supplemental doses of L-carnitine L-tartrate. J Strength Cond Res. (2007) 21(1):259-64.
103. Kraemer WJ et. al. Androgenic responses to resistance exercise: effects of feeding and L-carnitine. Med Sci Sports Exerc. (2006) 38(7):1288-96.
104. Calabrese V et. al. Acetylcarnitine and cellular stress response: roles in nutritional redox homeostasis and regulation of longevity genes. J Nutr Biochem. (2006) 17(2):73-88.
105. Ames BN and J Liu. Delaying the mitochondrial decay of aging with acetylcarnitine. Ann N Y Acad Sci. (2004) 1033:108-16
106. Mingrone G. Carnitine in type 2 diabetes. Ann N Y Acad Sci. (2004) 1033:99-107.
107. Lino A [Psycho-functional changes in attention and learning under the action of L-acetylcarnitine in 17 young subjects. A pilot study of its use in mental deterioration] Clin Ter. (1992) 140(6):569-73. [Article in Italian]
108. Genazzani AD et. al. Acetyl-l-carnitine as possible drug in the treatment of hypothalamic amenorrhea. Acta Obstet Gynecol Scand. (1991) 70(6):487-92.
109. Ames BN and Liu J. Delaying the mitochondrial decay of aging with acetylcarnitine. Ann N Y Acad Sci. (2004) 1033:108-16.
110. Begum G et. al. Physiological role of carnosine in contracting muscle. Int J Sport Nutr Exerc Metab. (2005) 15(5):493-514.
111. Mannion AF et. al. Carnosine and anserine concentrations in the quadriceps femoris muscle of healthy humans. Eur J Appl Physiol Occup Physiol. (1992) 64(1):47-50.
112. Suzuki Y High level of skeletal muscle carnosine contributes to the latter half of exercise performance during 30-s maximal cycle ergometer sprinting. Jpn J Physiol. (2002) 52(2):199-205.
113. Parkhouse WS et. al. Buffering capacity of deproteinized human vastus lateralis muscle. J Appl Physiol. (1985) 58(1):14-7.
114. Tallon MJ et. al. The carnosine content of vastus lateralis is elevated in resistance-trained bodybuilders. J Strength Cond Res. (2005) 19(4):725-9.
115. Gariballa SE, Sinclair AJ. Carnosine: physiological properties and therapeutic potential. Age Ageing. (2000) 29(3):207-10.
116. Dunnett M, Harris RC. Influence of oral beta-alanine and L-histidine supplementation on the carnosine content of the gluteus medius. Equine Vet J Suppl. (1999) 30:499-504.
117. Hill CA et. al. Influence of beta-alanine supplementation on skeletal muscle carnosine concentrations and high intensity cycling capacity. Amino Acids. (2007) 32(2):225-33.
118. Harris RC et. al. The absorption of orally supplied beta-alanine and its effect on muscle carnosine synthesis in human vastus lateralis. Amino Acids. (2006) 30(3):279-89

119. Stout JR et. al. Effects of twenty-eight days of beta-alanine and creatine monohydrate supplementation on the physical working capacity at neuromuscular fatigue threshold. J Strength Cond Res. (2006) 20(4):928-31.
120. Zoeller RF et. al. Effects of 28 days of beta-alanine and creatine monohydrate supplementation on aerobic power, ventilatory and lactate thresholds, and time to exhaustion. Amino Acids. (2006) Sep 5; [Epub ahead of print]
121. Hoffman J et. al. Effect of creatine and beta-alanine supplementation on performance and endocrine responses in strength/power athletes. Int J Sport Nutr Exerc Metab. (2006) 16(4):430-46.
122. Derave W Beta-alanine supplementation augments muscle carnosine content and attenuates fatigue during repeated isokinetic contraction bouts in trained sprinters. J Appl Physiol. (2007) Aug 9; [Epub ahead of print]
123. Camerino DC et. al. Taurine and skeletal muscle disorders. Neurochem Res (2004). 29: 135-142.
124 .Militante JD and JB Lombardini. Treatment of hypertension with oral taurine: experimental and clinical studies. Amino Acids. (2002) 23(4):381-93.
125. Franconi F, et. al. Taurine supplementation and diabetes mellitus. Curr Opin Clin Nutr Metab Care. (2006) 9(1):32-6.
126. Zhang M et. al. Beneficial effects of taurine on serum lipids in overweight or obese non-diabetic subjects. Amino Acids. (2004) 26(3):267-71.
127. Ward RJ et. al. Changes in plasma taurine levels after different endurance events. Amino Acids. (1999) 16(1):71-7.
128. Zhang M et. al. Role of taurine supplementation to prevent exercise-induced oxidative stress in healthy young men. Amino Acids. (2004) 26(2):203-7.
129. Cuisinier C. Changes in plasma and urinary taurine and amino acids in runners immediately and 24h after a marathon. Amino Acids. (2001) 20(1):13-23.
130. Baum M and M. WeiB. The influence of a taurine containing drink on cardiac parameters before and after exercise measured by electrocardiography. Amino Acids (2001) 20: 75-82.
131. K. -R. Geiß et. al. The effect of a taurine-containing drink on performance in 10 endurance-athletes Amino Acids (1994) 7: 45-56.
132. Alford C et. al. The effects of red bull energy drink on human performance and mood. Amino Acids. 2001;21(2):139-50.
133. Forbes SC et. al. Effect of Red Bull energy drink on repeated wingate cycle performance and bench-press muscle endurance. Int J Sports Nutr (2007) 17: 433-444.
134. Bendahan D, et. al. Citrulline/malate promotes aerobic energy production in human exercising muscle. Br J Sports Med. (2002) 36(4):282-9.
135. Callis A et. al. Activity of citrulline malate on acid-base balance and blood ammonia and amino acid levels. Study in the animal and in man. Arzneimittelforschung. (1991) 41(6):660-3.
136. Hickner RC et. al. L-citrulline reduces time to exhaustion and insulin response to a graded exercise test. Med Sci Sports Exerc. (2006) 38(4):660-6.
137. Romero MJ et. al. Therapeutic use of citrulline in cardiovascular disease. Cardiovasc Drug Rev. (2006) 24(3-4):275-90.
138. Rennie MJ. Claims for the anabolic effects of growth hormone: a case of the emperor's new clothes? Br J Sports Med. (2003) 37(2):100-5.
139. Collier SR, et. al. Growth hormone responses to varying doses of oral arginine. Growth Horm IGF Res. (2005) 15(2):136-9.
140. Lambert MI, et al. Failure of commercial oral amino acid supplements to increase serum growth hormone concentrations in male body-builders. Int J Sport Nutr. (1993) 3(3):298-305.
141. Fogelholm et. al. Low-dose amino acid supplementation: no effects on serum human growth hormone and insulin in male weightlifters. Int J Sport Nutr. (1993) 3(3):290-7.

142. Welbourne TC. Increased plasma bicarbonate and growth hormone after an oral glutamine load.Am J Clin Nutr. (1995) 61(5):1058-61.
143. Chromiak JA, Antonio J. Use of amino acids as growth hormone-releasing agents by athletes. Nutrition. (2002) 18(7-8):657-61.
144. Abel T et. al. Influence of chronic supplementation of arginine aspartate in endurance athletes on performance and substrate metabolism - a randomized, double-blind, placebo-controlled study. Int J Sports Med. (2005) 26(5):344-9.
145. Colombani et. al. Chronic arginine aspartate supplementation in runners reduces total plasma amino acid level at rest and during a marathon run. Eur J Nutr. (1999) 38(6):263-70.
146. Gravholt C et. al. Effects of a physiological GH pulse on interstitial glycerol in abdominal and femoral adipose tissue. Am J Physiol. (1999) 277(5 Pt 1):E848-54.
147. Samra JS Suppression of the nocturnal rise in growth hormone reduces subsequent lipolysis in subcutaneous adipose tissue. Eur J Clin Invest. (1999) 29(12):1045-52.
148. Douglas Paddon-Jones et. al. Potential Ergogenic Effects of Arginine and Creatine Supplementation J. Nutr. (2004) 134:2888S-2894S.
149. Bode-Boger SM. Effect of L-arginine supplementation on NO production in man. Eur J Clin Pharmacol. (2006) 62 Suppl 1:91-9.
150. McConell GK. Effects of L-arginine supplementation on exercise metabolism. Curr Opin Clin Nutr Metab Care. (2007) 10(1):46-51.
151. Clark MG et. al. Blood flow and muscle metabolism: a focus on insulin action. Am J Physiol Endocrinol Metab. (2003) 284(2):E241-58.
152. Marshall K. Therapeutic applications of whey protein. Altern Med Rev. 2004 Jun;9(2):136-56.
153. Lands LC et. al. Effect of supplementation with a cysteine donor on muscular performance. J Appl Physiol. (1999) 87(4):1381-5
154. Kinscherf R et. al. Low plasma glutamine in combination with high glutamate levels indicate risk for loss of body cell mass in healthy individuals: the effect of N-acetyl-cystein. J Mol Med (1996) 74: 393-400.
155. Sauvaire Y et. al.4-Hydroxyisoleucine: a novel amino acid potentiator of insulin secretion. Diabetes. (1998) 47(2):206-10.
156. Kochhar A, Nagi M. Effect of supplementation of traditional medicinal plants on blood glucose in non-insulin-dependent diabetics: a pilot study. J Med Food. (2005) 8(4):545-9.
157. Ruby BC et. al. The addition of fenugreek extract (Trigonella foenum-graecum) to glucose feeding increases muscle glycogen resynthesis after exercise. Amino Acids. (2005) 28(1):71-6.
158. Slivka D et. al. Glycogen resynthesis and exercise performance with the addition of fenugreek extract (4-hydroxyisoleucine) to post-exercise carbohydrate feeding.Amino Acids. (2007) Aug 21; [Epub ahead of print]
159. Playford RJ et. al. Colostrum and milk-derived peptide growth factors for the treatment of gastrointestinal disorders. Am J Clin Nutr. (2000) 72(1):5-14.
160. Mero A et. al.IGF-I, IgA, and IgG responses to bovine colostrum supplementation during training. J Appl Physiol. (2002) 93(2):732-9.
161. Buckley JD et. al. Effect of bovine colostrum on anaerobic exercise performance and plasma insulin-like growth factor I. J Sports Sci. (2003) 21(7):577-88.
162. Kuipers H et. al. Effects of oral bovine colostrum supplementation on serum insulin-like growth factor-I levels. Nutrition. (2002) 18(7-8):566-7.
163. Buckley JD et. al. Bovine colostrum supplementation during endurance running training improves recovery, but not performance.J Sci Med Sport. (2002) 5(2):65-79.
164. Hofman Z et. al. The effect of bovine colostrum supplementation on exercise performance in elite field hockey players. Int J Sport Nutr Exerc Metab. (2002) 12(4):461-9.

165. Antonio J et. al. The effects of bovine colostrum supplementation on body composition and exercise performance in active men and women. Nutrition. (2001) 17(3):243-7.
166. Brinkworth GD et. al. Oral bovine colostrum supplementation enhances buffer capacity but not rowing performance in elite female rowers. Int J Sport Nutr Exerc Metab. (2002) 12(3):349-65.
167. Brinkworth GD et. al. Effect of bovine colostrum supplementation on the composition of resistance trained and untrained limbs in healthy young men. Eur J Appl Physiol. (2004) 91(1):53-60.
168. Shing CM et. al. Effects of bovine colostrum supplementation on immune variables in highly trained cyclists. J Appl Physiol. (2007) 102(3):1113-22.
169. Shing CM et. al. The influence of bovine colostrum supplementation on exercise performance in highly trained cyclists. Br J Sports Med. (2006) 40(9):797-801.

<u>References cited for Chapter 13: Putting it all together</u>

1. Tipton KD and Witard OC. Protein requirements and recommendations for athletes: relevance of ivory tower arguments for practical recommendations. Clin Sports Med. (2007) 26(1):17-36.
2. MacDougall JD et. al. The time course for elevated muscle protein synthesis following heavy resistance exercise. Can J Appl Physiol. (1995) 20(4):480-6.
3. Chesley A et. al. Changes in human muscle protein synthesis after resistance exercise. J Appl Physiol. (1992) 73(4):1383-8.
4. Economos CD et. al. Nutritional practices of elite athletes. Practical recommendations. Sports Med. (1993) 16(6):381-99.
5. Elliott MC et. al. Power athletes and distance training: physiological and biomechanical rationale for change Sports Med. (2007) 37(1):47-57.
6. Manore MM. Dietary recommendations and athletic menstrual dysfunction. Sports Med. (2002) 32(14):887-901.
7. Venderley AM and Campbell WW. Vegetarian diets: nutritional considerations for athletes. Sports Med. (2006) 36(4):293-305.

Index